Industrial Applications of Soil Microbes

Volume 1

Edited by

Ashutosh Gupta

Shampi Jain

&

Neeraj Verma

Department of Agriculture Science
AKS University, Satna
Madhya Pradesh–485001
India

Industrial Applications of Soil Microbes

(Volume 1)

Editors: Ashutosh Gupta, Shampi Jain and Neeraj Verma

ISBN (Online): 978-981-5039-95-5

ISBN (Print): 978-981-5039-96-2

ISBN (Paperback): 978-981-5039-97-9

Published by Bentham Science Publishers Pte. Ltd. Singapore. All Rights Reserved.

First published in 2022.

need for a court order if at any point you breach any terms of this License Agreement. In no event will any delay or failure by Bentham Science Publishers in enforcing your compliance with this License Agreement constitute a waiver of any of its rights.

3. You acknowledge that you have read this License Agreement, and agree to be bound by its terms and conditions. To the extent that any other terms and conditions presented on any website of Bentham Science Publishers conflict with, or are inconsistent with, the terms and conditions set out in this License Agreement, you acknowledge that the terms and conditions set out in this License Agreement shall prevail.

Bentham Science Publishers Pte. Ltd.
80 Robinson Road #02-00
Singapore 068898
Singapore
Email: subscriptions@benthamscience.net

BENTHAM SCIENCE

CONTENTS

FOREWORD

I am indeed very happy that this book on the scientific and industrial aspects of soil microbes is coming up. Since the discovery of microorganisms like bacteria, fungi, mycorrhiza, *etc.*, and their association with plants, scientists have been attracted to the microbial association with plants and their role in the ecosystem. Studies on plant-associated soil microbes and their use in the industry have been a fascinating area of research for the last two decades and also play a crucial role in plant growth and development, such as nutrient cycling, crop productivity, and nitrogen fixation. Nowadays, scientists and industries are exploring the use of microbes in the fields of agricultural and industrial production of secondary metabolites from these beneficial microorganisms. Several bio-products are available on the market, and many more are likely to come in the future.

This book, entitled **"Industrial Applications of Soil Microbes"**, Volume I, is a praiseworthy step towards popularizing the industrial applications of microbes inhabiting the soil environment. This volume contains fifteen contributed articles on various aspects of soil microbes, their industrial utility, and their role in improving soil quality and yield. The various chapters of the present book have been written by reputed scientists in the microbiology field. It will provide the most recent information on the uses of microorganisms and their applications in industry and agriculture in one place.

The authors have covered most of the interesting topics, like preliminary information about soil microbes to highly advanced technologies like degradation of inorganic wastes, bioremediation, nutrient transformation, and the role of nanotechnology in agriculture. Some information about IPR, its rules, and regulations is also discussed. The latest references used by the authors will also provide the most recent developments in the industrial/agricultural uses of soil microbes.

I do not doubt that this book will be a significant scientific contribution and a source of inspiration to students and scientists who are trying to understand the mechanism of interaction of microbes with soil as well as their industrial applications.

Abdul Samad
Ex-Chief Scientist & Acting Director
CSIR-Central Institute of Medicinal
and Aromatic Plants (CSIR-CIMAP)
Lucknow- 226015, India

PREFACE

The rapid increase in the population, industrialization, and so many human activities are all interrelated with each other and cause harm not only to the environment but also to the soil ecosystem. The soil ecosystem comprises various microorganisms, which may be beneficial or harmful to plants, soil, animals, and human beings. Some microbes, such as algae, bacteria, fungi, protozoans, and mycorrhizae, inhabiting soil are highly important for living beings, including humans.

These soil microbes carry out different biochemical reactions, leading to beneficial products. The ability of soil microbes to decompose organic waste is one of the major important uses. Beneficial soil microbes are being utilized in various chemical reactions in industries to produce products, which can be directly used in agriculture to increase the yield of the crops.

Indiscriminate use of fertilizer in the soil, barren land, and various pollutants present in the atmosphere results in an unhealthy environment for living beings. Either using the soil microbes in modified or unmodified ways may enhance the production of agriproducts and cause a healthy environment. The beneficial soil microorganisms as such or their products can be used in agriculture to get a better yield. This will certainly be beneficial to the farmers.

Sometimes a crazy idea also means a new school of opinion. We have been emphasizing here the soil microbes and their uses in agriculture, and we found that the information is scanty and scattered about the soil microbes and their uses. Thus, students and researchers could not get all this information in one place. Therefore, we decided to come up with a book entitled **"Industrial Applications of Soil Microbes"**, which carries the latest information on the subject.

In this volume of the book, an attempt has been made to present the basic and applied aspects of soil microbes with the latest information about microbial diversity in soil, their involvement in different processes either directly or indirectly for enhancement of crop yield, maintaining the ecosystem, reducing the pollution, future applications of soil microorganisms as nanoparticles, intellectual property rights, and their management.

Although there are other textbooks on soil microbes, it is hoped that the present book will be highly helpful to the students, academicians, scientists, and entrepreneurs who are working in this field. Further, it may provide information to the growers of the crops, giving them some principled knowledge to enhance their crop productivity.

The editors will be pleased to receive comments on the style and content of this volume for inclusion in further editions.

Ashutosh Gupta

Shampi Jain

&
Neeraj Verma
Department of Agriculture Science
AKS University
Satna, Madhya Pradesh
– 485001, India

ACKNOWLEDGEMENT

The authors are highly thankful to the researchers, authors, and editors who contributed their research in the form of chapters for this book. We are also thankful to the Hon'ble Chancellor, Mr. B.P. Soni and Pro-Chancellor, Er. Anant Soni, AKS University, Satna, MP, India for permitting us and providing the facilities and encouragement. The authors are highly thankful to our colleagues for their constructive suggestions. We are also thankful to Mr. Ashish Khare, Designer, AKS University, for editing the figures. We can not forget the help of our family members. Without their love, affection, and encouragement, we could not have completed this work.

The authors express thanks to Bentham Science Publishers, for allowing this project and especially to Humaira Hashmi for timely help and guidance.

Ashutosh Gupta

Shampi Jain

&
Neeraj Verma
Department of Agriculture Science
AKS University
Satna, Madhya Pradesh
− 485001, India

List of Contributors

A.U. Siddiqui	Pacific College of Agriculture, Pacific University, Udaipur (Rajasthan), India
Alka Rani	ICAR-Indian Institute of Soil Science, Nabibagh, Berasia Road, Bhopal, India
Anandkumar Naorem	Indian Council of Agricultural Research-Central Arid Zone Research Institute, Regional Research Station-Bhuj, Gujarat-370105, India
Anurag Singh	Department of Microbiology, Ram Lal Anand College, University of Delhi, Benito Juarez Road, New Delhi-110021, India
Arbind Kumar Gupta	Department of Soil Science, College of Forestry, Banda University of Agriculture & Technology, Banda-210001 U.P., India
Ashutosh Kumar	Department of Agriculture Economics, Faculty of Agriculture Science and Technology, AKS University, Satna, M.P., India
Ashwini A. Waoo	Department of Biotechnology, Faculty of Life Sciences and Technology, AKS University, Satna, MP, India
Atik Ahamad	KVK, Bharari, Jhansi, U.P., India
Avinash Pandey	ICAR-Indian Institute of Agricultural Biotechnology, Ranchi, 834010, India
Bineeta Devi	Department of Genetics and Plant Breeding, Faculty of Agriculture Science and Technology, AKS University, Satna, M.P., India
B.J. Yogesh	Department of Microbiology, The Oxford College of Science, Bangalore, India
B.S. Chandrawat	SKN College of Agriculture, SKNAU, Jobner (Rajasthan), India
Deo Kumar	Department of Soil Science & Agril Chemistry, College of Agriculture, Banda University of Agriculture & Technology, Banda-210001, U.P., India
F. Ahmad	JC Bose University of Science and Technology, YMCA, Faridabad, Haryana, India
Harshraj Kanwar	Rajasthan Agricultural Research Institute, Durgapura, SKNAU, Jobner (Rajasthan), India
Jaison Maverick	Department of Agricultural chemistry and Soil Science, Bidhan Chandra Krishi Viswa Vidyalaya, Mohanpur, West Bengal-741252, India
Jitendra Kumar	ICAR-Indian Institute of Soil Science, Nabibagh, Berasia Road, Bhopal, India
Kishor Chand Kumhar	Deen Dayal Upadhyay Centre of Excellence for Organic Farming, CCS HAU, Hisar – 125004, India
K.R. Maurya	Department of Horticulture, Faculty of Agriculture Science and Technology, AKS University, Satna, M.P., India
K.S. Karthika	ICAR -NBSS & LUP, Regional Centre, Bangalore-560 024, India
M. Mohanty	ICAR-Indian Institute of Soil Science, Nabibagh, Berasia Road, Bhopal, India
Nidhi S. Chandra	Department of Microbiology, Ram Lal Anand College, University of Delhi, Benito Juarez Road, New Delhi-110021, India
Nishant K. Sinha	ICAR-Indian Institute of Soil Science, Nabibagh, Berasia Road, Bhopal, India
P.B. Khaire	Department of Plant Pathology and Agril. Microbiology, PGI, MPKV, Rahuri, India

Parul Sundha	ICAR-Central Soil Salinity Research Institute, Karnal-132001, Haryana, India
Pooja Verma	ICAR-Central Soil Salinity Research Institute, Karnal-132001, Haryana, India
Prabha Susan Philip	ICAR -NBSS & LUP, Regional Centre, Delhi-110012, India
Priya Bhatia	Department of Microbiology, Ram Lal Anand College, University of Delhi, Benito Juarez Road, New Delhi-110021, India
Priyanka Chandra	ICAR-Central Soil Salinity Research Institute, Karnal-132001, Haryana, India
Raj Kumar	Division of Crop Protection, ICAR-CPCRI, Kasaragod, Kerala, India
Ramesh Nath Gupta	Department of Plant Pathology, Bihar Agricultural University, Sabour-813210, India
Rashid Pervez	Division of Nematology, ICAR-Indian Agricultural Research Institute, New Delhi, India
Rinki	ICAR-Indian Institute of Wheat and Barley Research, Karnal-132001, Haryana, India
R.S. Chaudhary	ICAR-Indian Institute of Soil Science, Nabibagh, Berasia Road, Bhopal, India
S. Ahmad	JC Bose University of Science and Technology, YMCA, Faridabad, Haryana, India
S. Bharathi	Department of Microbiology, The Oxford College of Science, Bangalore, India
S.V. Pawar	Department of Plant Pathology, Sumitomo Chemical India Limited, Mumbai, India
Sachin Patel	Indian Council of Agricultural Research-Central Arid Zone Research Institute, Regional Research Station-Bhuj, Gujarat-370105, India
Sanjay Gupta	Department of Biotechnology, Jaypee Institute of Information Technology, Sector 62, Noida, UP-201309, India
Savitha Santosh	ICAR-Central Institute for Cotton Research, Nagpur, Maharashtra - 440010, India
Shiva Kumar Udayana	Krishi Vigyan Kendra, Dr. YSRHU, Venkataramannagudem, West Godavari, Andhra Pradesh-534101, India
Shivangi Agnihotri	Department of Biotechnology, Faculty of Life Sciences and Technology, AKS University, Satna, MP, India
Shreya Kapoor	Department of Microbiology, Ram Lal Anand College, University of Delhi, Benito Juarez Road, New Delhi-110021, India
Sneha S. Nair	Department of Agricultural Microbiology, College of Horticulture, Kerala Agricultural University, Thrissur- 680656, Kerala, India
Someshree S. Mane	Department of Plant Pathology and Agril. Microbiology, PGI, MPKV, Rahuri, India
Vandana Gupta	Department of Microbiology, Ram Lal Anand College, University of Delhi, Benito Juarez Road, New Delhi-110021, India
Vanita Pandey	ICAR-Indian Institute of Wheat and Barley Research, Karnal-132001, Haryana, India

CHAPTER 1

Soil Bacteria- Our Allies in Building Soil Health

Sneha S. Nair[1], Prabha Susan Philip[2,*] and K.S. Karthika[3]

[1] *Department of Agricultural Microbiology, College of Horticulture, Kerala Agricultural University, Thrissur-680656, Kerala, India*

[2] *ICAR -NBSS & LUP, Regional Centre, Delhi-110012, India*

[3] *ICAR -NBSS & LUP, Regional Centre, Bangalore-560 024, India*

Abstract: Microorganisms give life to the soil and provide a variety of ecosystem services to plants. Soil bacteria are the strongest candidates for determining soil health. Bacterial communities are important for the health and productivity of soil ecosystems. Therefore, we must have thorough knowledge of the diversity, habitat, and ecosystem functions of bacteria. In this chapter, we will discuss the functional, metabolic, and phylogenetic diversity of soil bacteria and highlight the role of bacteria in the cycling of major biological elements (C, N, P, and S), detoxification of common soil pollutants, disease suppression, and soil aggregation. This chapter also underlines the use of soil bacteria as indicators of soil health. We have concluded the chapter by taking note of the present agricultural practices that call for concern regarding the natural soil microflora and steps to return biological activity to the soil.

Keywords: Bioremediation, Disease suppression, Nutrient cycling, Soil health.

INTRODUCTION

Soil is a unique product formed by a combination of geological parent material, soil biota activity, land use pattern, and disturbance regimes. It is a niche for a diversity of microorganisms such as bacteria, archaea, fungi, insects, annelids, and other invertebrates, as well as plants and algae. Among various microbes present in the soil, bacteria are the most abundant. They provide a multitude of ecosystem services in the soil, such as nutrient cycling, disease suppression, soil aggregation, bioremediation, *etc.* The most critical function of bacteria in soil is the regulation of biogeochemical transformations. Soil is a complex medium rich in a variety of minerals and complex organic molecules.

A wide variety of bacteria, especially those belonging to actinobacteria and proteobacteria, can degrade complex organic molecules such as cellulose, hemice-

* **Corresponding author Prabha Susan Philip:** ICAR -NBSS & LUP, Regional Centre, Delhi-110012, India; E-mail: prabha0825@gmail.com

Ashutosh Gupta, Shampi Jain & Neeraj Verma (Eds.)

llulose, lignin, and chitin to simple sugars and amino acids, thereby leading to the formation and turnover of soil organic matter (OM) that includes mineralization and sequestration of carbon (C). The entire process of nitrogen fixation into the rhizosphere or plant compartment is mediated by soil bacteria. They also significantly contribute to phosphorus, sulphur, iron, and magnesium pools in the soil. Soil bacteria are also active agents of bioremediation and disease suppression. They use complex mechanisms such as biosorption, bioaccumulation, biotransformation, bioprecipitation, and bioleaching to reduce toxic metals in soil. Soil aggregate formation is a major function of these soil dwellers. They do so by producing extracellular polymeric substances (EPS), which act as gelling materials and help stick the aggregates together. This, in turn, helps in increasing the water-holding capacity of the soil. As the key producers of different enzymes in the soil, such as dehydrogenases, catalases, phosphatases, *etc.*, soil bacteria are also commonly used as indicator organisms to assess soil health.

Bacteria serve numerous roles in the pedosphere. Being able to survive even in the extremes of the environment, they have become strong candidates for building soil health. For that reason, it is necessary to understand their soil habitat, diversity, and the ecosystem services they provide. Therefore, in this chapter, we will briefly consider diversity, abundance, distribution, and, in particular, their role in regulating ecosystem services.

SOIL AS A CULTURE MEDIUM FOR BACTERIA

Soil is the outer loose material of the earth's surface, which provides mechanical support and essential nutrients to plants. The nutrients in the soil are obtained by physical, biological, and biochemical reactions related to the decomposition of organic weathering of the parent rock. The soil also acts as a niche for many microorganisms, especially bacteria, which are the most abundant microorganisms in the soil, providing them with sufficient nutrients and a beneficial microclimate for their survival. The diversity, abundance, and activity of bacterial communities are dependent on soil depth, soil pH, soil temperature, soil nutrients, soil moisture, *etc*.

Soil Depth

Soil is divided into different horizons based on its physical, chemical, and biological characteristics. Typical horizons that develop in agricultural soils include the A, B, and C horizons. Bacteria exist in all the horizons, but their abundance and diversity are highest in the top 10 cm of soil and decline with depth. According to Aislabie and coworkers [1], the A horizon has the highest density of soil microbes and plant roots because of its high organic matter content and is considered to be the site of humification in soil. As water infiltrates the A horizon, organic compounds and minerals like iron, aluminium, clay, *etc.*, get leached out to the B horizon. The B horizon (subsoil) retains non-mobile constituents and thus tends to be enriched in minerals such as quartz. Below the B horizon are the partially weathered stones, forming a C horizon above the solid, unweathered bedrock. Fierer and coworkers [2] discovered that the maximum microbial biomass exists in the top 0–5 cm of soil depth. The reason for this might be the abundant availability of organic carbon in the A horizon. Bacteria that can grow facultatively with the help of inorganic ions mostly occupy the deeper horizons of soil. As one moves to the C horizon, nutrients, as well as water and oxygen, become deficient, leading to the habitation of certain extremophiles.

Soil pH

Soil pH also influences the occurrence and distribution of bacteria as it determines the chemical activity of protons in soil, which are the foundation for all enzymatic reactions, mineralization and precipitation, surface complexation, and other geochemical reactions in soil [3]. Based on optimal growth pH, bacteria can be distinguished into three groups: acidophiles that grow best at pH < 5, neutrophiles that grow optimally at pH between 5 and 9, and alkaliphiles that grow fastest above pH 9 (Table 1). For a microbe with a pH range spanning 4 pH units, assuming that its optimal pH is near the middle point of the pH range, a deviation of one unit from this optimal pH can reduce its growth rate by about 50% [4]. Therefore, soil pH is an important parameter that determines the bacterial population in the soil. Most soil bacteria show optimum growth at neutrophilic pH, but some bacterial species can thrive well under extremes of pH.

Table 1. Examples of bacteria surviving in extreme habitats [5].

Parameter	Type	Examples
Oxygen Concentration	Obligate aerobes	*Azotobacter, Rhizobium, Micrococcus, Streptomyces coelicolor*, and *Nitrobacter*
	Obligate anaerobes	Bacteroidetes, *Clostridium, Methanobacterium*, and sulphate reducers
	Facultative anaerobes	*Escherichia coli, Paenibacillus*, and *Bacillus*
	Aerotolerant	*Streptococcus* sp.
	Microaerophilic	*Azospirillum* and *Campylobacter*

(Table 1) cont.....

Parameter	Type	Examples
Temperature	Thermophilic	*Geobacillus, Ureibacillus,* and *Brevibacillus* [6]
	Hyperthermophile	*Pyrococcus, Thermococcus,* and *Desulfurococcus* [7]
	Psychrophile	*Methylobacter, Methylosinus, Acidobacteria* sp., and *Actinobacteria* sp.
	Psychrotroph	*Rhodococcus* sp [8], *Azospirillum himalayense* [9], and *Flavobacterium phocarum* [10]
Nutrition	Photoautotrophs	Cyanobacteria
	Photoheterotrophs	*Heliobacterium mobilis* and *Heliobacterium modesticaldum*
	Chemoautotrophs /chemolithotroph	Alcaligenes, *Acidithiobacillus, Methanococcus,* and *Thiobacillus*
	Chemoheterotroph/chemoorganotroph	*Azotobacter, Azomonas, Azospirillum, Klebsiella, Clostridium, Desulfotomaculum,* and *Methylomonas*
pH	Acidophiles	*Thiobacillus ferrooxidans, T. thiooxidans, Sulfobacillus acidophilus, Acidimicrobium ferrooxidans, Acidiphilium* sp., and *Alicyclobacillus* sp. [11]
	Alkaliphiles	*Bacillus alcalophilus, B. agaradherens, B. clarkii, B. clasusii, B. gibsonii, B. haemophilus, B. halodurans, B. Horikoshii,* and *B. pseudofirmus* [12]

Soil Nutrients

Soil is biologically alive as it contains sufficient nutrients for microbial growth. Oxygen and hydrogen are abundantly present in the upper layers of soil. Nitrogen is abundant in the atmosphere as nitrogen gas but is unavailable to bacteria except to the few species of nitrogen-fixing bacteria, as rainwater fixes very little nitrogen as nitrates in the soil. The ultimate source of carbon in soil is carbon dioxide. Based on the nutrient requirements, bacteria in soil can be classified as phototrophs and chemotrophs. Photoautotrophs are organisms that carry out photosynthesis. Using energy from sunlight, carbon dioxide and water are converted into organic materials to be used in cellular functions such as biosynthesis and respiration. Photoheterotrophs obtain their energy from sunlight and carbon from organic material, not carbon dioxide. They can be contrasted with chemotrophs that obtain their energy from the oxidation of electron donors in their environments. Chemoautotrophs can synthesize their organic molecules from the fixation of carbon dioxide. The energy required for this process comes from the oxidation of inorganic molecules such as iron, sulfur, or magnesium. Chemoheterotrophs, unlike chemoautotrophs, are unable to synthesize their organic molecules. Instead, they consume carbon molecules synthesized by other organisms. They do so, however, while still obtaining energy from the oxidation of inorganic molecules like the chemoautotrophs. Chemoheterotrophs can only thrive in environments that are capable of sustaining other forms of life due to their dependence on these organisms for carbon sources (Table **1**).

Soil Moisture

The moisture content of the soil is correlated with the oxygen status. Soil water content is important in regulating oxygen diffusion, with maximum aerobic microbial activity occurring at moisture levels between 50% and 70% of the water holding capacity [5]. A high moisture content (metric potential > −0.01 MPa) can result in a low oxygen supply, leading to a decrease in the rate of organic matter decomposition. A low moisture content can decrease microbial activity, as water is an essential nutrient for the activation of most of the microbial enzymes. Although some bacteria can survive in extreme moisture conditions such as drought and waterlogged soils, the majority of soil bacteria prefer optimum moisture conditions of 50-70% water holding capacity. Drought-tolerant bacteria in soil include *Pseudomonas*, *Enterobacter*, *Pantoea*, *Klebsiella*, *Arthrobacter,* and *Ochrobactrum*. Whereas methanogens, denitrifiers, sulphate reducers, *etc.*, dominate waterlogged soil [13].

Soil Aeration

Oxygen is an essential micronutrient for most of the bacteria in the soil as it acts as the terminal electron acceptor in the electron transport chain. Bacteria that cannot survive in the absence of oxygen are called obligate aerobes. They contain enzymes that can neutralize free radicals and peroxides of oxygen, such as catalase, peroxidase, and superoxide dismutase (SOD). On the other hand, bacteria that are killed in the presence of oxygen are called obligate anaerobes. They may live by fermentation, anaerobic respiration, bacterial photosynthesis, or methanogenesis. Facultative anaerobes are organisms that can switch between aerobic and anaerobic types of metabolism. Under anaerobic conditions, they grow by fermentation or anaerobic respiration, but in the presence of O_2, they switch to aerobic respiration. Two more categories of bacteria exist in soils that can tolerate low levels of oxygen. They are aerotolerant anaerobes and microaerophiles. Aerotolerant anaerobes are indifferent to the presence of oxygen. They do not use oxygen because they usually have a fermentative metabolism, but they are not harmed by the presence of oxygen, as obligate anaerobes are, whereas microaerophiles require oxygen to survive, but require environments containing lower levels of oxygen than those that are present in the atmosphere (*i.e.* <21% O_2; typically 2–10% O_2). Obligate anaerobes usually lack all three enzymes. Aerotolerant anaerobes do have SOD but no catalase.

Soil Temperature

Bacteria are known to live at extremes of temperature (Table **1**). Temperature has a direct effect on bacteria, as most enzymes get denatured at very high temperatures. Organisms have a specific range of temperature at which their biological activity operates, which is defined by cardinal points (minimum, optimum, and maximum temperature). Soil bacteria may be subdivided into several subclasses based on their cardinal points for growth. Bacteria with an optimum temperature (T) near 37°C are called mesophiles. Thermophiles are bacteria that thrive at temperatures ranging from 45 to 70°C. Bacteria (mostly Archaea) with an optimum T of 80 to 115°C are referred to as hyperthermophiles. Thermophiles possess several heat-stable enzymes wherein they showcase different critical amino acid substitutions at a few locations in the enzyme that allow the protein to fold in a unique, heat-stable way. Besides enzymes and other macromolecules in the cell, the cytoplasmic membranes of thermophiles and hyperthermophiles are counteracted by constructing membranes with more long-chain and saturated fatty acid content and lower unsaturated fatty acid content than is found in the cytoplasmic membranes of mesophiles. Thermophiles also have a high G+C content in their DNA, such that the melting point of the DNA (the temperature at which the strands of the double helix separate) is at least as high as the organism's maximum T [5]. The cold-loving bacteria, on the other hand, are psychrophiles, defined by their ability to grow at 0°C. A variant of a psychrophile (which usually has an optimum temperature of 10-15°C) is the psychrotroph, which grows at 0°C but displays an optimum temperature in the mesophile range, nearer room temperature. Psychrophiles have unsaturated fatty acids in their plasma membrane, which remain liquid at low temperatures but can also be denatured at moderate temperatures. Psychrophiles also have enzymes with a greater content of α-helix and a lesser content of β-sheet, which helps them function at temperatures of at or near 0 degrees.

The soil environment is very complex and provides diverse microbial habitats. It differs with regard to climate, vegetation, location, *etc*. Despite these factors, we now know that bacteria can survive harsh environments. Biological activity is an important fraction of soil, and the microbes that colonize these soils are influenced by the soil pH, moisture, temperature, aeration, *etc*. A combination of these factors influences the microbial community of any soil.

CLASSIFICATION OF SOIL BACTERIA

Bacteria are prokaryotic (without a true nucleus), single-celled organisms varying in size and shape. A single gram of soil may contain 10^3 to 10^6 unique species of bacteria [14]. Soil bacteria can be classified based on their shapes (rod, sphere, spiral and pleomorphic), size (varying from 0.5μm), cell wall structure (Gram-positive and Gram-negative), oxygen requirements (obligate aerobes, facultative anaerobes, microaerophilic and obligate anaerobes), nutritional differences (autochthonous and zymogenous) and growth and reproduction (autotrophic, chemotrophic and heterotrophic). With new advances in DNA sequencing, scientists have classified bacteria into closely related phyla. DNA sequencing technology can be used to study relationships between unknown and known organisms and provides an estimate of the genetic diversity of organisms in a given community. In this section, we will go through important bacterial classifications in soil microbiology.

Classification Based on Nutritional Differences

Soil bacteria are most commonly classified based on nutritional differences into autochthonous and zymogenous [15]. Autochthonous microbes are indigenously present in soil and have adaptive mechanisms to survive in natural conditions with extremely low nutrient supplies (oligotrophic environment) [16]. They are ecologically defined as k- strategists. In contrast, zymogenous bacteria, or r-strategists consist of actively fermenting forms, which require nutrients that are quickly exhausted for their activity. Under nutrient-sufficient conditions, these microbes increase to great numbers and then return to detectable numbers as nutrients diminish. Major differences between the two groups are listed in Table 2. R- and k-strategists have often been described in relation to ecological succession, with r-strategists being abundant in early successional stages and k-strategists prevailing in late successional stages [17]. In a nutshell, r-strategists are adapted to maximize their intrinsic rate of growth when resources are abundant, while k-strategists are adapted to compete and survive when populations are near carrying capacity and resources are limited [18].

Table 2. Major difference between k and r strategists [19].

Characteristics	k-strategists/Autochthonous	r-strategists/Zymogenous
Growth rate	Low	High
Energy supply	Scarce	Plentiful
Nutrient competitive ability	Low	High
Survival potential	High	Low
Preferred ecosystem type	Mature	Young

(Table 2) cont.....

Characteristics	k-strategists/Autochthonous	r-strategists/Zymogenous
Persistency of colonization	High	Low
Examples	*E. coli* and *Caulobacter*	*Pseudomonas aeruginosa, Nitrosomonas,* and *Methylomonas*

Molecular Classification of Soil Bacteria

Soil bacteria were previously classified by studying individual strains through cultivation in laboratories. However, this approach neglects the vast majority of soil bacteria that are unculturable. Recently, molecular approaches have gained importance as a useful tool for the characterization of bacterial communities [20]. The estimated relative abundance of the major phyla varies between different soils (or samples). Jansen [21] identified the dominant soil bacterial taxa in libraries of 16S rRNA and 16S rRNA genes and highlighted members of the phyla Proteobacteria, Acidobacteria, and Actinobacteria to be widespread and often abundant, whereas members of the Verrucomicrobia, Bacteroidetes, Firmicutes, Chloroflexi, Planctomycetes, and Gemmatimonadetes to be less prevalent (Table 3). The Proteobacteria are a metabolically diverse group of organisms in several subphyla, four of which, α-, β-, γ-, and δ-Proteobacteria are abundantly present in the soil. They play an important role in determining soil nutrient characteristics, as most of the bacteria in this phylum are involved in different biogeochemical cycles. These proteobacteria can be either r- or k-strategists. Adding low molecular weight carbon to soil increased the relative abundances of β- and γ-Proteobacteria while spiking soils with recalcitrant carbon increased the relative abundance of α-, β-, and δ-Proteobacteria [22].

Table 3. Molecular classification of soil bacteria [5].

Phyla/Subphyla	Examples
α-Proteobacteria	*Sphingomonas, Rhizobium, Mesorhizobium, Bradyrhizobium, Methylobacter, Methylophilus, Nitrospira, Nitrobacter,* and *Rhodobacter*
β Proteobacteria	*Burkholderia, Alcaligenes, Acidovorax, Collimonas, Nitrosospira, Thiobacillus, Rhodocyclus,* and *Methylomonas*
γ-Proteobacteria	*Pseudomonas, Xanthomonas, Azotobacter, Thiocapsa,* and *Chromatium*
δ-Proteobacteria	*Desulfovibrio* and *Bdellovibrio*
ε-Proteobacteria	*Helicobacter* and *Campylobacter*
Acidobacteria	*Acidobacterium*
Actinobacteria	*Arthrobacter, Rhodococcus, Streptomyces, Mycobacterium, Rubrobacter, Terrabacter,* and *Acidimicrobium*
Verrucomicrobia	*Chthoniobacter* and *Opitutus*

(Table 3) cont.....

Phyla/Subphyla	Examples
Bacteroidetes	*Chitinophaga*
Firmicutes	*Clostridium, Bacillus*, and *Lactobacillus*
Gemmatimonadetes	*Gemmatimonas*

Acidobacteria is a group of bacteria similar to Proteobacteria in their metabolic and phylogenetic diversity, ubiquitous nature, and abundance in soil habitats. Genome sequencing of three-cultured soil Acidobacteria suggests that bacteria belonging to this phyla may be oligotrophs that metabolize a wide range of simple and complex carbon sources [23]. They also appear well suited to low nutrient conditions, tolerate fluctuations in soil moisture, and are capable of nitrate and nitrite reduction, but not denitrification or nitrogen fixation.

Actinobacteria and Firmicutes contribute as major Gram-positive bacteria in the soil. Actinobacteria are aerobic and spore-forming bacteria, belonging to the order Actinomycetales, with most of them characterized by substrate and aerial mycelium growth. They are also responsible for the characteristically "earthy" smell of freshly turned healthy soil. The phyla include soil actinobacteria (such as *Frankia*), which live symbiotically with the plants whose roots pervade the soil, fixing nitrogen; *Streptomyces*, known for their ability to produce antimicrobial compounds; and members of the genus *Mycobacterium*, which are important pathogens. They are metabolically diverse aerobic heterotrophs, playing an important role in the cycling of carbon, nitrogen, phosphorus, potassium, and several other elements in the soil [24]. The relative abundance of Actinobacteridae in the soil increases following the addition of labile carbon sources [22].

Firmicutes include heat-stable endospore formers in soil and are aerobic to facultatively anaerobic (genus *Bacillus*) and anaerobic (genus *Clostridium*). *Bacillus* species are known for their plant growth-promoting activity, besides their antagonistic activity against many pathogens. The most abundant bacilli in soil include *Bacillus subtilis*, *B. cereus*, *B. thuringiensis*, *B. pumilus*, and *B. Megaterium*.

Bacteroidetes are another group of bacteria present in the soil. They are involved in the aerobic degradation of complex organic molecules such as starch, proteins, cellulose, and chitin. They are Gram-negative, non-spore-forming, anaerobic or aerobic, and rod-shaped bacteria, whereas Verrucomicrobia is characterized as an unculturable soil microbiota. The most abundant verrucomicrobial class, Spartobacteria, contains free-living taxa. Verrucomicrobial isolates display a wide variety of features, such as methanotrophy, nitrogen fixation, pathogenicity, *etc.*

Another ubiquitous organism on the list is Planctomycetes. Examples of these include *Singulisphaera acidiphila*, *S. rosea*, and *Paludisphaera borealis*, which are psychrotolerant and can grow at low temperatures, down to 4–6°C [25]. They are most commonly known in soil for their ability to perform anaerobic ammonium oxidation (anammox) [26]. Planctomycetes possess distinctive phenotypic features that are highly unusual among bacteria. These include the lack of peptidoglycan in their protein cell walls (as reproduction mostly by budding process), resistance to antibiotics that inhibit cell wall synthesis, such as the β-lactam and vancomycin (due to lack of peptidoglycan) and synthesis of sterols [27].

Classification Based on Soil Functions

Bacteria perform many important ecosystem services in the soil, such as improved soil structure, soil aggregation, nutrient cycling, *etc.* (Table **4**). Even though phylogenetic classification is the most accurate method of classification, as soil microbiologists, we need to ponder upon a much more diverse division of soil bacteria as they have numerous ecosystem functions in the soil. Therefore, the whole chapter about soil bacteria will be dealt with keeping in mind the ecosystem functions of the bacteria in the soil.

Table 4. Roles of soil microbes in soil ecosystems.

Role	Description	Example	References
Decomposers	Organic matter decomposition, including cellulose, hemicellulose and lignin	*Bacillus subtilis, B. pumilus, Pseudomonas fluorescens*, and *Streptomyces viridosporus*	[28]
Nitrogen Fixers	Nitrogen fixation	Alpha and Beta-proteobacteria (*Azospirillum, Azorhizobium, Bradyrhizobium, Methylobacterium, Burkholderia*), Firmicutes (*e.g. Paenibacillus, Heliobacterium*), and Cyanobacteria	[29]
P Solubilizers	Capable of solubilizing and mineralizing insoluble soil phosphorus for the growth of plants	*Pseudomonas* spp., *Agrobacterium* spp., and *Bacillus circulans*	[30]
K Solubilizers	Releasing K from inorganic and insoluble pools of total soil K through solubilization	*Delftia acidovorans, Paenibacillus macerans, Pantoea agglomerans, Pseudomonas* spp., *P. aureofaciens, P. chlororaphis, P. fluorescens, P. solanacearum, Bacillus mucilaginous, B. pumilus, B. subtilis, B. amyloliquefaciens, B. firmus, B. licheniformis*, and *B. megaterium*	[31]

(Table 4) cont.....

Role	Description	Example	References
Plant growth promoters	Production of growth hormones	*Azotobacter* spp., *Rhizobium* spp., *Pantoea agglomerans, Rhodospirillum rubrum, Pseudomonas fluorescens, Bacillus subtilis,* and *Paenibacillus polymyxa*	[31]
	Abiotic stress tolerance to plants through increased proline production, ACC deaminase activity, induction of antioxidant enzymes, and enhanced nutrient uptake	*Pseudomonas syringae, Pseudomonas fluorescens, Enterobacter aerogenes* (Salt tolerance), *Arthrobacter* sp., *Bacillus* sp. (osmotic stress), *Methylobacterium oryzae,* and *Burkholderia* sp. (heavy metal toxicity)	[32 - 34]
Disease suppressors	Siderophore production, enzyme production, and antibiotic production	*Bacillus subtilis* and *Pseudomonas* sp.	[35, 36]
Modify soil structure	Production of EPS and biofilms	*Gluconacetobacter xylinus, Agrobacterium tumefaciens, Rhizobium* spp., and *Alcaligenes faecalis*	[37, 38]

ROLE OF BACTERIA IN SUSTAINING SOIL HEALTH

The soil microbial community is fundamental in any ecosystem because it takes part in organic matter decomposition and nutrient cycling, influencing the chemical and physical properties of the soil and, consequently, primary productivity. In the subsequent sections, we will discuss in detail the ecosystem services that soil bacteria provide.

Nutrient Cycling

Nutrient cycling is the whole process of use, movement, and recycling of nutrients within the ecosystem and is one of the most important ecosystem functions that soil bacteria execute. Most of the elements on earth are nutritionally essential and are prerequisites for living organisms to complete their life cycle. For the growth and development of higher plants, 17 nutrients are considered essential as per criteria developed by Arnon and Stout [39]. These include both macro and micronutrients derived mainly from the atmosphere- carbon (C), hydrogen (H), oxygen (O); primary nutrients- nitrogen (N), phosphorus (P), potassium (K); secondary nutrients- calcium (Ca), magnesium (Mg), sulphur (S); and micronutrients- iron (Fe), manganese (Mn), zinc (Zn), copper (Cu), boron (B), molybdenum (Mo), chlorine (Cl) and nickel (Ni). All these nutrients play a key role in a plant's life cycle. Therefore, these nutrients must be cycled in a way to make them available for plant uptake. Nutrient cycles are complex processes involving biological, geological, and chemical processes, and for this reason, the term "biogeochemical cycle" is often used.

These circuits are completed and connected by various nutrient transformation steps wherein some or whole steps are microbe mediated. Nutrient transformation is a key process that makes nutrients available for plant uptake. Microorganisms, especially bacteria, provide a notable contribution to nutrient cycling. As a part of this chapter here, we discuss four important nutrient cycles, which play a significant role in soil ecosystem balance.

Carbon Cycle

Carbon is the structural component of organic molecules like carbohydrates, proteins, and lipids that build living tissues. It is the backbone element for all life forms in the biosphere. The element is of great relevance as the major greenhouse gases that regulate the pace of global warming and climate change are carbon compounds; carbon dioxide (CO_2) and methane (CH_4). Organic compounds released in the form of root exudates, lysates, secretions, and other organic residues serve as the energy sources for belowground microfauna, especially the soil bacterial community.

The global carbon cycle describes the complex transformations and carbon fluxes between the major components of the Earth's system. Carbon is stored both as organic and inorganic compounds in four major Earth reservoirs, including the atmosphere, lithosphere, biosphere, and hydrosphere [40]. Two important processes of the C cycle are photosynthesis and respiration. Through the process of photosynthesis, CO_2 marks its entry into the biosphere, and through respiration it exits and leaves back into the atmosphere. The C cycle begins with biological C fixation, *i.e.*, CO_2 from the atmosphere is fixed as complex organic molecules in the terrestrial system (plants and microorganisms) by the process of photosynthesis. The processed form of C by photoautotrophs is utilized by heterotrophs as their energy source.

Carbon is released back into the atmosphere from the terrestrial ecosystem through several processes, like the burning of fossil fuels and crop residues, the decomposition of dead and decaying organic matter, and respiration by living organisms and roots. Decomposition by saprophytes (bacteria and fungi) is the breakdown of dead and decaying organic residue and the release of carbon and other nutrients back into the environment, and it is one of the most important roles of bacteria. By way of the decomposition process, nutrients or minerals are released back into the soil system; hence, this process is often referred to as mineralization.

The C:N ratio is a fundamental indicator of biogeochemical cycles in ecosystems. Any shift in C:N stoichiometry could pose great impacts on the nutrient cycling and the composition and structure of plant communities, affecting ecosystem service functions at local, regional, national, and global scales [41]. Carbon is important for bacteria as it is an energy-producing factor, and nitrogen for building microbial body tissues. There is always a narrowing of the carbon-nitrogen ratio when organic matter decomposes. Bacteria maintain a C: N ratio of 4:1 to 8:1, less compared to other microorganisms. This stoichiometry is important as it determines the direction of nutrient fluxes in the soil system.

Methane production, or methanogenesis, is also a crucial step in the carbon cycle. Reduced compounds like methane accumulate in certain anaerobic environments when CO_2 is used as a terminal electron acceptor in anaerobic respiration by archaea called methanogens. Methanogens in both natural anaerobic soil and aquatic environments lead to methane accumulation and cause serious environmental concern as CH_4 is a greenhouse gas involved in global warming. Biological oxidation of this methane in the soil can be carried out by methanotrophs, thereby reducing the atmospheric methane levels.

N-Cycle and Microbial Nitrogen Transformations

The N-cycle begins with the inert dinitrogen (N_2) form and follows it through the processes of fixation, mineralization, immobilization, nitrification, ammonia volatilization, leaching, runoff, plant assimilation, and completion with denitrification. Microorganisms, especially soil bacteria, are involved in all major aspects of the cycle. Plant available N in soil depends on the balance between rates of mineralization, nitrification, and denitrification. On earth, N is distributed in two large pools – as an inert form of atmospheric molecular nitrogen and a biologically reactive form of nitrogen, namely, NO_3, NH_4, and organic nitrogen. These two large pools are interconnected and controlled by microbially mediated processes of nitrogen fixation and denitrification [42]. The entire process of fixing N_2 into the rhizosphere or plant compartment and the transformation of one form into the other comprises the nitrogen cycle.

N Fixation

The N-cycle starts with biological N fixation, wherein gaseous inert N from the atmosphere is made available to plants by bacteria such as *Rhizobium* (in legumes). Microbial mediated nitrogen fixation is one of the prime mechanisms contributing to large additions of mineral nitrogen into the soil ecosystem. In 1888, Dutch microbiologist Martinus Beijerinck was the first to identify *Rhizobium*, a class of bacteria for its ability to fix elemental N from the atmosphere into the roots of leguminous plants by forming nodules. Several scientific reports disclose the roles of bacterial members from the families Azotobacteriaceae, Spirillaceae, Enterobacteriaceae, Bacillaceae, Pseudomonadaceae, and Achromobacteriaceae in nitrogen fixation. As plants cannot take up N in its atmospheric form, it needs to be transformed to either as nitrate (NO_3^-) or ammoniacal (NH_4^+) form to facilitate its uptake by plants [43]. Biological N fixation includes the processes through which atmospheric N is converted to ammonia with the involvement of nitrogenase enzymes by symbiotic N fixers, associative N fixers, and free-living N fixers [44].

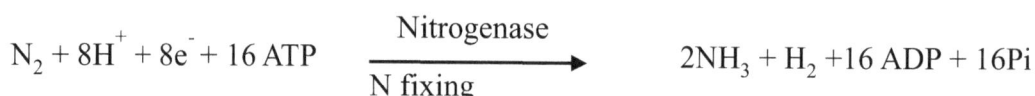

$$N_2 + 8H^+ + 8e^- + 16\,ATP \xrightarrow[\text{N fixing}]{\text{Nitrogenase}} 2NH_3 + H_2 + 16\,ADP + 16Pi$$

Symbiotic N Fixation

Symbiotic N fixation is a well-known example of the symbiotic association that exists between the roots of leguminous crops and N-fixing bacteria. This results from the complex interaction between the host plant and rhizobia (*Rhizobium, Bradyrhizobium, Sinorhizobium,* and *Mesorhizobium*) [45]. *Rhizobium* and *Bradyrhizobium* belonging to α proteobacteria are the main N fixers that inhabit the root nodules of leguminous crops.

Non-Symbiotic N Fixation or Associative N Fixation

The method of N fixation also termed associated N fixation refers to the process in which bacteria fix nitrogen by using carbon compounds as an energy source, and bacteria release the fixed nitrogen at the root surface and in the cellular interstices of root tissue only after lysis of the bacterial cells [46, 47]. *Azotobacter, Azospirillum,* and *Enterobacter* are major soil bacteria involved in non-symbiotic N fixation. Associative N fixation in plant root interiors has gained high significance and agronomic importance in sugarcane, sweet potato, and rice, *etc.*

Free Living Nitrogen Fixation

In free-living N fixation, the fixation of N occurs without a formal microbe plant symbiotic relationship. The process is ubiquitous in terrestrial ecosystems and the major free-living N fixers include many bacterial phyla such as Alphaproteobacteria (*Bradyrhizobium* and *Rhodobacter*), Betaproteobacteria (*Burkholderia* and *Nitrosospira*), Gammaproteobacteria (*Pseudomonas* and *Xanthomonas*), deltaproteobacteria, firmicutes, cyanobacteria, *etc.*

N Mineralization

The majority share of N in the soil is present as complex organic molecules, which need to be converted into plant-usable forms by microorganisms present at the soil-root interface. Mineralization of organic nitrogen is a three-step process, namely aminization, ammonification, and nitrification. These processes are governed by the carbon: nitrogen (C: N) ratio of the organic substrate present. In the case of fresh and easily decomposable organic material with a narrower C: N ratio (for legumes, 30:1), mineralization takes place faster, whereas for high molecular weight compounds like lignin and humus with a wider C: N ratio (for sawdust, 600:1), mineralization is very slow [48].

Aminization

Heterotrophic bacteria, actinomycetes, and fungi are capable of enzymatically digesting proteins and other proteinaceous compounds released from roots or organic residues into amino acids and amines, and the process is known as aminization.

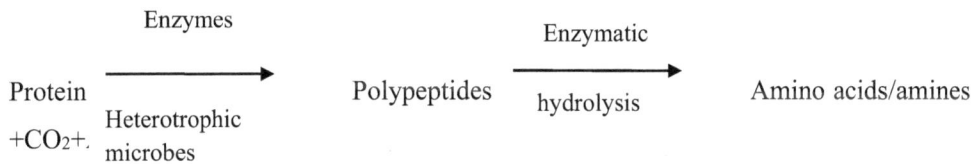

$$\text{Protein} + CO_2 + \xrightarrow[\text{Heterotrophic microbes}]{\text{Enzymes}} \text{Polypeptides} \xrightarrow[\text{hydrolysis}]{\text{Enzymatic}} \text{Amino acids/amines}$$

The amino acids so produced are either utilized by microbes (immobilization) or mineralized to NH_3 by ammonification. Amino acids act as an important source of N uptake for plants under some circumstances, and the existence of amino acid uptake systems in plant roots has also been reported.

Ammonification

In this process, amino acids or amines released by root exudation are transformed into ammonia by ammonifiers (bacteria, fungi, and actinomycetes) using enzymes.

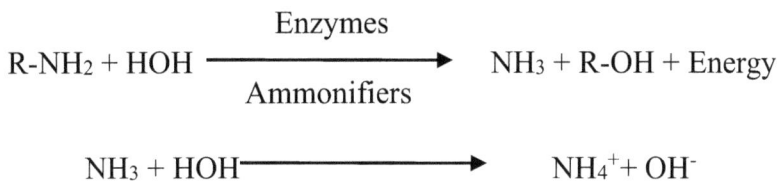

$$R\text{-}NH_2 + HOH \xrightarrow[\text{Ammonifiers}]{\text{Enzymes}} NH_3 + R\text{-}OH + \text{Energy}$$

$$NH_3 + HOH \longrightarrow NH_4^+ + OH^-$$

Nitrification

This process involves two-step oxidation processes wherein the oxidation of ammonium ion (NH_4^+) is converted to nitrate ion (NO_3^-) by nitrifying organisms. In the first step, NH_4^+ is converted to nitrite (NO_2^-) by *Nitrosomonas*, *Micrococcus*, and *Nitrospira*, and then the conversion of NO_2^- to NO_3^- is mediated by *Nitrobacter* and *Nitrocystis*.

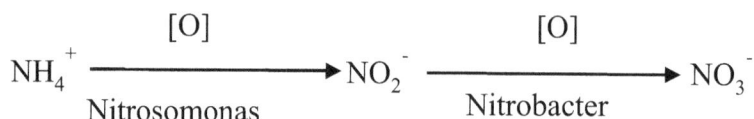

$$NH_4^+ \xrightarrow[\text{Nitrosomonas}]{[O]} NO_2^- \xrightarrow[\text{Nitrobacter}]{[O]} NO_3^-$$

Immobilization

Soil microorganisms utilize inorganic N forms for building up their body tissues, thereby leading to the temporary unavailability of nitrogen for plant uptake. This process of conversion of inorganic nitrogen into organic forms is termed immobilization. It is a reversal of the mineralization process. The main factor governing the immobilization process is the C: N ratio of the substrate available for bacteria. The more complex the organic substrate, the wider the C: N ratio, leading to a higher N demand by saprophytes feeding on it. To meet their elevated N demand, decomposers utilize either added or native N sources immediately available to them. The process of immobilization leads only to the short-term unavailability of N for plants.

Denitrification

The process of denitrification mainly happens under anoxic conditions in which bacteria of the genera *Thiobacillus denitrificans*, *Thiobacillus thioparus*, *Pseudomonas*, *Micrococcus*, *Achromobacter*, and *Bacillus* reduce NO_2^- and NO_3^- leading to the release of gases NO, N_2O, and N_2 back into the atmosphere.

These microorganisms use organic compounds as their energy source, which are available in plenty in the soil. Favourable conditions for the denitrification process include limited O_2 supply, high concentration of NO_3^-, presence of soil moisture, carbohydrates source and warm temperatures [49, 50]. Nitrate and nitrite act as electron acceptors instead of O_2 for respiration by microorganisms in denitrification.

$$NO_3^- \longrightarrow NO_2^- \longrightarrow NO \longrightarrow N_2O\!\uparrow \longrightarrow N_2\!\uparrow$$

Losses of mineral N from terrestrial ecosystems are mainly through microbial-mediated processes of nitrification and denitrification. Dissimilatory nitrate reduction (DNRA) is a process mediated by bacteria and fungi wherein the respiratory reduction of nitrate to a more stable ammonium form takes place, thereby limiting losses of N from the soil system.

Phosphorus Cycle

Phosphorus (P) is a primary nutrient element vital for plant growth and development. It plays an important role in energy transfer, root formation and growth, photosynthesis, the translocation of sugars and starches, *etc*. Microorganisms, specifically bacteria, are integral to phosphorus transformation in soil and play a key role in mediating its availability to plants. Knowledge regarding the microbial contribution and opportunities for manipulation of the specific microbial communities to enhance P availability and use efficiency is important as global phosphorus reserves are limited and getting exhausted day by day. The best option available is to increase P use efficiency by employing a specific microbial community capable of doing so [51].

Phosphorus in the soil solution is mainly brought about by the weathering of minerals (apatites), the application of phosphatic fertilizers, and organic manure addition. The added P in the solution gets transformed through adsorption and precipitation processes into P products of varying degrees of solubility. Soil bacteria are capable of mineralizing a fraction of such chemically immobilized P and bringing it to the soil solution pool for plant uptake. During this process, a portion of it gets tied up in microbial tissues for a short period, which upon decaying of microbial tissues gets released into soil solution [48].

Bacteria secrete organic acids like lactic and glycolic acids that aid in the solubilization of immobilized inorganic phosphates, thereby enhancing P availability to plants. P solubilizing bacteria are cultured and available for commercial application as biofertilizers or as plant-growth promoting bacteria (PGPRs). Several *Pseudomonas* spp. are also associated with P mobilization [52].

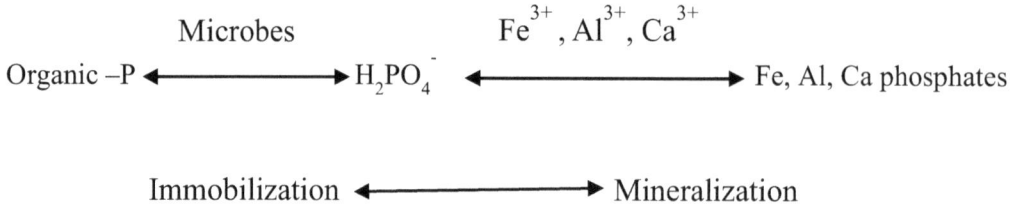

$$\text{Organic } -P \xleftrightarrow{\quad \text{Microbes} \quad} H_2PO_4^- \xleftrightarrow{\quad Fe^{3+}, Al^{3+}, Ca^{3+} \quad} \text{Fe, Al, Ca phosphates}$$

$$\text{Immobilization} \longleftrightarrow \text{Mineralization}$$

C:P Ratio

Net mineralization of organic P happens if the organic residue added to soil has a carbon to phosphorus (C:P) ratio below 200:1, while net immobilization of soluble P occurs at a C:P ratio above 300:1.

Sulphur Cycle

Sulphur (S) is the 10th most abundant element in the universe and the sixth most abundant element in microbial biomass [53]. It is a constituent element of many vitamins, proteins, and hormones that play critical roles in both climates and the health of various ecosystems. The majority of the Earth's sulphur is stored underground in rocks and minerals as sulphate salts buried deep within ocean sediments. The transformation and fate of this element in the environment are critically reliant on microbial activities.

The S cycle is both atmospheric and terrestrial. Similar to N, the majority of S exists as organic sulphur compounds. Since elemental S has the ability to exist in a wide range of stable redox states, it is involved in various oxidation and reduction biochemical reactions. Within the terrestrial portion, the cycle begins with the weathering of rocks, S then comes into contact with air where it is converted into sulphate (SO_4). Organic S compounds upon mineralization under aerobic upland conditions to SO_4 by the action of S oxidizing bacteria, *Thiobacillus*.

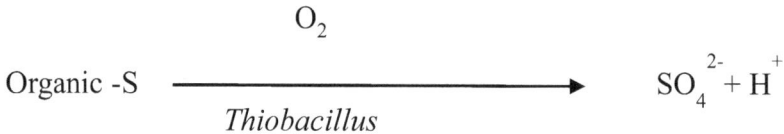

$$\text{Organic -S} \xrightarrow[\textit{Thiobacillus}]{\quad O_2 \quad} SO_4^{2-} + H^+$$

The sulphate is taken up by plants and microorganisms and converted into organic forms; animals then consume these organic forms through the foods they eat, thereby moving the sulphur through the food chain. The addition of bulky organic manures low in S content leads to a microbial proliferation, which tends to meet their S requirement from inorganic SO_4-S, leading to immobilization and rendering S temporarily unavailable to plants. Later, as these microbes die, microbial S is again mineralized to provide the plant with usable SO_4. Under reduced conditions, SO_4^{2-} ions are reduced by anaerobic bacteria, *Desulphovibrio* and *Desulphotomaculum*, to sulphite and sulphide, respectively. In the redox process, the SO_4^{2-} ion is the last one to undergo reduction because of its low redox potential value of -215 mV. There are also a variety of natural sources that emit S directly into the atmosphere, including volcanic eruptions, the breakdown of organic matter in swamps and tidal flats, and the evaporation of water. Besides this, SO_2, as a part of industrial pollution, also reaches the atmosphere and is brought back to the atmosphere in the form of acid rain. Thus, S eventually cycles back into the Earth or comes down in rainfall. A continuous loss of sulphur from terrestrial ecosystem runoff occurs through drainage into lakes, streams, and eventually oceans.

C:S Ratio

S in the soil is associated with organic carbon in a fixed carbon to sulphur (C:S) ratio and maintenance of the C:N:S ratio at favourable levels of about 125:140:1 is essential to ensure S availability to plants. Otherwise available S in the soil, which is subjected to various microbial transformations, may lead to temporary unavailability to crops.

RECYCLING WASTES AND DETOXIFICATION

Worldwide, contamination of soil due to natural geogenic sources such as volcanic releases, forest fires, erosion and anthropogenic activities such as mining, industrial activities, activities linked to transportation in and around urban centers, waste and sewage disposal, warfare, agricultural and livestock activities, *etc.*, has resulted in an increased release of a wide range of xenobiotics. The main source of soil pollution is anthropogenic [54]. The major pollutants in soil include heavy metals, nitrogen, phosphorus, pesticides, polycyclic aromatic hydrocarbons, persistent organic pollutants (POPs), per- and polyfluorinated alkyl substances (PFAS) and radionuclides. Soil pollution is an alarming issue as it threatens food security both by reducing crop yields due to toxic levels of contaminants and by causing the produced crops to be unsafe for consumption [55]. Unhealthy soil affects crop production as well as human and animal health through food contamination. Therefore, there is an urgent need to clean up these pollutants for the betterment of soil, human and animal health. Many methods of soil remediation have been developed over the years, among which bioremediation is an option that offers the possibility to destroy or render harmless various contaminants using natural biological activity [56].

Bioremediation

Bioremediation is the use of living organisms (primarily microorganisms) to degrade environmental pollutants into less toxic forms. It either uses the naturally occurring microflora to degrade the contaminants or microorganisms may be isolated from elsewhere and brought to the contaminated site. Among all the soil microorganisms, bacteria are a strong candidate in bioremediation owing to their survival in the extremes of the environment.

Types of Bioremediation

Bioremediation can be categorized into *ex situ* and *in situ* based on the place where wastes are removed. *Ex Situ* bioremediation takes place somewhere away from the contamination site and therefore requires transportation of contaminated soil or pumping of groundwater to the site of bioremediation. The types include biopiles, windrows, bioreactors, and land farming. A biopile is an above-ground pile of excavated polluted soil, followed by nutrient amendment and sometimes aeration to enhance bioremediation by basically increasing microbial activity. Windrower is similar to biopile with the difference of periodic turning of piled polluted soil in the former, which hastens the bioremediation process [57]. Bioreactors involve treating contaminated soils in a vessel, where the contaminants are converted to the specific product(s) following a series of biological reactions. This method has the additional advantage of mimicking and maintaining the natural environment to provide optimum growth conditions for bacteria that degrade the pollutants. Among all the *ex situ* methods, land farming is the simplest bioremediation technique. Here, soils are excavated and/or tilled and are carefully applied on a fixed layer of support above the ground surface to allow aerobic biodegradation of pollutants by autochthonous microorganisms [58].

The *in situ* methods include biosurping, bioventing, biosparging, and natural attenuation. Here the contaminants are treated at their site of formation. The most common among them is biosparging and bioventing. Bioventing involves controlled stimulation of airflow by delivering oxygen and adding nutrients to the unsaturated (vadose) zone to increase bioremediation by increasing the activities of indigenous microbes. Biosparging is similar to bioventing however, unlike bioventing, the air is injected at the saturated zone, which can cause upward movement of volatile organic compounds to the unsaturated zone to promote biodegradation. *In situ* bioremediation is a superior method for the cleaning of contaminated environments because it saves transportation costs and uses harmless microorganisms to eliminate chemical contamination.

Bacterial Roles in Bioremediation

Heavy Metal Removal

Bacteria possess the ability to remove heavy metals from the soil through biosorption, bioaccumulation, biotransformation, bioprecipitation, and bioleaching. Biosorption is an energy-independent process mediated by an interaction between positively charged metal ions and negatively charged bacterial exopolysaccharides (EPS) on the cell surface of bacteria through complexing, coordination, physical adsorption, chelation, ion exchange, inorganic precipitation, and/or a combination of these processes [59]. Several anionic ligands such as carboxyl, amine, hydroxyl, phosphate, and sulfhydryl groups are located on the bacterial cell, which aid in this process. Biosorption is a reversible process carried out by living as well as dead cells.

On the contrary, bioaccumulation is a process of active transport and extracellular (periplasm) or intracellular metal accumulation, where metals are precipitated, methylated, or bound to compounds and metal-binding proteins inside the microbial cells. It is an active process by living cells in which the removal of heavy metals depends on the metabolic activity of a living organism. Once inside the cells, the toxic heavy metals can be converted from one oxidation state to another (biotransformation), hence reducing their harmfulness. These metals, in their oxidized form, could serve as terminal electron acceptors during the anaerobic respiration of bacteria. They can also be precipitated or crystallized inside or outside the cell, which causes the transformation of metal into a sparingly soluble form, which lowers their toxicity at the same time. Bioleaching is the application of microorganisms for the transformation of compounds of metals present in the environment in the form of sparingly soluble substances (most often sulfides) into forms easily soluble where the removal of metals is an easy task [60]. For any bioremediation process to work, microorganisms with a proven ability to remediate and tolerate heavy toxicity should be utilized.

Pesticide Detoxification

Pesticides are chemical substances used to kill or reduce pest populations to a tolerable level. Most of the pesticides in soil are recalcitrant in nature, polluting soils, water, and air, raising serious concern for human health. Therefore, it is necessary to degrade these pesticides for the safety of our ecosystem. Apart from biological degradation, pesticides in the soil can be degraded by physical or chemical methods, but these methods are not preferred due to the development of secondary pollutants [61].

Bioremediation of pesticides by biological systems includes biodegradation and biotransformation. These two words are faintly different from each other. Biodegradation involves the biological synonyms and reactions where the chemical structure of the pesticide is modified, thereby reducing its toxicity, whereas biotransformation involves modification as well as translocation of the pesticides. Biodegradation is generally recognized as the biggest contributor to degradation. Plants, animals, and fungi typically transform pesticides for detoxification through metabolism by enzymes, whereas bacteria more commonly metabolize them, ultimately decomposing them into some small molecules, such as CO_2 and H_2O.

The whole degradation mechanism in bacteria is divided into three parts. The target is initially adsorbed to the cell through the surface of the cell membrane, following which the pesticide goes through the cell membrane and is enzymatically broken down in the membrane *via* oxidation, reduction, and hydrolysis [62]. Further degradation of the pesticide is carried out through mineralization and co-metabolism. Mineralization involves the conversion of organic compounds into inorganic compounds under the action of soil microbes. They could be used as a source of microbial nutrients and then be degraded to inorganic matters like carbon dioxide and water by microorganisms. Co-metabolic refers to some chemical substances like insecticides, fungicides, and herbicides that are not degraded by bacteria or fungi easily, but only by adding some organic matter such as exogenous or iso-biomass as the primary energy source [63].

Polycyclic Aromatic Hydrocarbons (PAHs)

Polycyclic aromatic hydrocarbons (PAH) are produced mostly by the incomplete combustion of organic materials at very high temperatures, such as coal, tar, gas, automobile exhaust, *etc.*, during industrial or anthropogenic activities or as a result of geothermal reactions like the production of fossil fuels and minerals. PAHs are considered important soil pollutants due to their electrochemical stability, persistency, resistance to biodegradation, and carcinogenicity in humans [64]. The PAHs have a natural potential for bioaccumulation in various food chains, which makes their presence in the environment quite alarming [65] and, therefore, they are a potential human hazard.

Some bacteria in soil can degrade the PAHs. They use aerobic or anaerobic pathways to degrade them. In the aerobic catabolism of aromatics, oxygen plays the dual role of being a terminal electron acceptor as well as acting as a co-substrate for the hydroxylation and oxygenolytic ring cleavage of the aromatic ring of PAHs. In contrast, anaerobic catabolism uses reductive reactions wherein anaerobes use nitrate, sulfate or ferric ions as alternative final electron acceptors

[66]. Bacteria principally favour aerobic degradation of PAHs. The first step of aerobic catabolism is the hydroxylation of the aromatic ring of PAH *via* dioxygenases, resulting in the formation of a *cis*-dihydrodiol, which gets rearomatized to a diol intermediate by the action of a dehydrogenase. The diol intermediates are then cleaved by dioxygenases either through an ortho or meta-cleavage pathway, leading to intermediates such as catechols that are ultimately converted to TCA cycle intermediates [64, 67]. Of all the known soil bacteria that degrade PAHs, *Pseudomonas putida* is a strong candidate.

Disease Suppression

The soil is a niche for a vast majority of microorganisms, including beneficial microbes as well as soil-borne pathogens. Many pathogens can even form survival structures under extreme environmental conditions and survive as a source of infection for the vegetation. However, biotic and abiotic factors can contribute to the irreversible loss of viability of pathogenic propagules. In this section, we will discuss the biotic mode of disease suppression.

Disease suppressive soils are soils in which the pathogen does not establish or persist, establishes but causes little or no damage, or establishes and causes disease for a while but thereafter the disease is less important, although the pathogen may persist in the soil [68]. Disease suppressiveness may be of two types; general and specific suppressiveness. General suppression of soils is attributed to the overall effect of the microbial community, principally through resource competition, whereas specific suppression is due to the concerted activities of specific groups of microorganisms antagonistic to the pathogen during some stage of its life cycle. Specific suppressiveness of soils can, in contrast to general suppressiveness, be transferred to conducive soils by mixing small amounts (1–10% w/w) of the suppressive soil into the conducive soil [69].

Soil suppressiveness encompasses various mechanisms, including competition, antibiosis, allelopathy, hyperparasitism, and the induction of plant disease resistance.

Competition

Microorganisms require the same nutrients, be it the beneficial microflora or the pathogens, for their survival in their soil habitat. The disease-suppressing bacteria affect the growth of pathogens by competing for the same mineral nutrients in the niche. These bacteria are equipped with better nutrient absorption mechanisms such as siderophore production, establishing partial sinks for nutrients, altering the physical habitat such as pH, temperature, and aeration where pathogens cannot survive. Microbial decomposers also produce inhibitors during competitive interactions, which not only affect the direct microbial competitors but also harm other soil-inhabiting organisms [70].

Antibiosis

Specific or nonspecific microbial metabolites, lytic agents, enzymes, volatile compounds, or other microbial toxins are produced by bacteria as agents of disease suppression. Antibiotic production is one of the most effective ways by which bacteria suppress pathogens in the soil. Antibacterial, antifungal, and antinematode activity has been identified in the antibiotic-producing strains of different plant growth-promoting rhizobacteria (PGPR) such as 2,4-diacetylphloroglucinol, phenazine-1-carboxyclic acid, phenazine-1-carboxamide, pyoluteorin, pyrrolnitrin, oomycin A, butyrolactones, kanosamine, zwittermicin-A, aerugine, rhamnolipids, cepaciamide A, pseudomonic acid, azomycin, antitumor antibiotic FR901463, cepafungins, and antiviral antibiotic karalicin [71]. The inhibitory action of each of these antibiotics may be different, such as inhibiting cell wall formation, protein synthesis, and other cellular constituents. Among the antibiotics with a role in disease suppressive soils, 2,4-diacetylphloroglucinol (DAPG) and phenazines (PHZ) have been widely studied. DAPG and PHZ producing fluorescent *Pseudomonas* species are known to suppress take-all disease of wheat caused *by Gaeumannomyces graminis* var. *tritici* [36], whereas DAPG and pyrrolnitrin were shown to be involved in the suppression of *R. solani* [72]. *Bacillus subtilis* AU 195 is another potent bacterial strain isolated from soil that produces bacillomycin D and is known to control the growth of aflatoxin-producing *Aspergillus flavus* [35]. Another mode of action through bacterial antibiosis is through the production of lytic enzymes such as hydrolase, laminarinase, and protease. The activity of these enzymes includes degradation of fungal spores, lysis of mycelium and breakdown of hyphal contents [73]. Volatile compounds with antimicrobial activities have also been proposed to play a role in the disease suppressiveness of soils. Other microbial by-products that may also contribute to pathogen suppression include hydrogen cyanide (HCN) produced by fluorescent dyes, which blocks the cytochromeoxidase pathway in pathogens.

Hyperparasitism

Hyperparasitism is the direct competitive interaction between two parasites in which one organism is gaining nutrients from the other. This kind of interaction is often observed between fungi. In bacteria, *Bdellovibrio bacteriovorus* is the only known predatory bacterium, which has the unusual property of using the cytoplasm of other gram-negative bacteria as nutrients. Predation by *B. bacteriovorus* strains of other plant pathogenic bacteria such as *Agrobacterium tumefaciens*, *Xanthomonas vesicatoria*, *X. campestris* pv. *campestris*, *Erwinia carotovora* pv. *carotovora*, *Pseudomonas syringae* pv. *glycinea*, *P. syringae* pv. *tomato*, *P. marginalis*, and *Erwinia herbicola* have been confirmed in similar tests [74].

Induced Resistance

Induced resistance is a physiological state of enhanced defensive capacity of the plant triggered by biological or chemical inducers, which protects plant tissues not exposed to the initial attack against future attacks by pathogens and herbivorous insects [75]. Induced resistance may be of two types- (a) SAR (Systemic acquired resistance), triggered by plant pathogens, and (b) ISR (Induced systemic resistance), triggered by root-colonizing mutualistic microbes. The inducible resistance in plants to an array of pathogens is called systemic acquired resistance (SAR). Salicylic acid (SA) is a key compound in SAR and is frequently produced after pathogen infection that leads to the expression of pathogenesis-related (PR) proteins. Some of these PRs are 1,3-glucanases and chitinases, which are capable of hydrolyzing fungal cell walls [76].

ISR is triggered in plants by beneficial microbes through elicitor molecules such as microbe-associated molecular patterns (bacterial flagellin, chitin, and lipopolysaccharides), volatile organic compounds (VOCs), or siderophores. These elicitor molecules have been found to trigger drastic changes in plant growth patterns, generally by altering hormone signalling [77]. Beneficial ISR-eliciting microbes do not directly activate defense responses but sensitize the whole plant for faster and stronger activation of defense responses upon invasion by pathogens [78].

Soil Aggregate Formation

Soil aggregates are groups of soil particles that bind to each other more strongly than to adjacent particles. The space between the aggregates provides pore space for the retention and exchange of air and water. Therefore, soil aggregates affect the soil's bulk properties, including organic carbon content, water content, and niche availability. Aggregates can be broadly classified into macroaggregates (>250 µm) and microaggregates (20-250 µm). Microaggregates help bind soil organic carbon and protect it from removal by erosion, and macroaggregates limit oxygen diffusion and regulate water flow [79]. Soil aggregate is formed of sand, silt, clay, organic matter, root hairs, microorganisms and their "glue" like secretions, mucilages, extracellular polysaccharides, and hyphae, as well as the resulting pores. Soil bacteria live within microaggregates as well as macroaggregates. Some cells are entrapped within the mineral matrix during aggregate formation, whereas some colonize the aggregate exterior [80]. Other than stabilizing microbial members and enhancing community interactions, aggregates can also change community function and microbial traits through spatial confinement [81]. For example, under desiccation or environmental stress, nitrifying bacteria persist within the protected interior of macroaggregates [82].

Soil bacteria form aggregates by producing organic compounds called extracellular polymeric substances (EPS), which act as a gelling material and help stick the aggregates together. Microbial EPS can enhance the aggregation of soil particles and benefit plants by trapping nutrients and maintaining the moisture of the environment. Water trapped in the EPS matrix provides a highly hydrated environment that dries more slowly than its surroundings and therefore buffers the biofilm cells against fluctuations in water potential and protects the cells from desiccation [83]. In addition, EPSs have unique characteristics, such as biocompatibility, gelling, and thickening capabilities, with industrial applications [84].

EPS include highly hydrated polymers that are mainly composed of water, polysaccharides, proteins and DNA, which are fundamental for microbial life and provide an ideal environment for chemical reactions, nutrient entrapment and protection against environmental stresses such as salinity and drought. The exopolysaccharides secreted by the bacterial community include alginate from *P. aeruginosa*, cellulose formed by *Gluconacetobacter xylinus*, *Agrobacterium tumefaciens*, *Rhizobium* spp., succinoglycan (EPS I) and galactoglucan (EPS II) produced by several rhizobial strains [37], levan produced by *Bacillus subtilis* and certain *Pseudomonas* sp [85], curdlan produced by *Alcaligenes faecalis* [38], *etc.* Extracellular proteins in the EPS are composed of enzymes and structural proteins such as lectins, extracellular polysaccharides, proteins, and DNA. These are highly hydrated hydrophilic molecules, but EPS molecules have hydrophobic properties. This is due to the presence of surfactants and lipids such as surfactin, viscosin, and emulsan. Other major functions of the EPS include aiding in microbial adhesion to surfaces and microbe-host interactions such as symbiosis, acting as carbon reserves and aiding in extracellular digestion, and protection of the bacterial community from abiotic (drought, salinity, temperature extremes) and biotic (antimicrobial) stress.

Aggregate stability is a critical function of any soil. Unstable aggregates at the surface of the soil can be easily broken down, sealing the pores and thereby reducing water infiltration into the soil, resulting in runoff. When erosion occurs, soil particles, dissolved nutrients, and organic matter particles are removed from agricultural fields, leading to loss of fertility and depletion of organic matter. Field management, including reduced tillage operations, the addition of organic amendments and bio-fertilizers, and planting and harvesting methods, can impact both aggregate size distribution and stability.

Indicators of Soil Health

Soil health is influenced by several factors, which can be broadly classified as biotic and abiotic factors. Biotic factors include soil microorganisms, macro- and mesofauna, their biomass, activity, enzymes released during their reactions, soil respiration, and the role of soil microbes in organic matter decomposition or soil nutrient cycles [86]. Soil microbial communities' responses are an indicator of soil health. Soil bacteria act as indicators of soil health by acting as 1) decomposers by being involved in organic matter decomposition, breaking down pesticides and pollutants in soil, 2) mutualists by forming relationships with plants, such as in the case of N fixing bacteria, 3) disease suppressors that possess anti-fungal activity and enhance plant growth, and 4) lithotrophs or chemoautotrophs, in which a few are involved in N cycling and degradation of pollutants.

INDICATORS OF SOIL HEALTH BY BACTERIA

Indicators of Soil Biodiversity

Genetic, functional, and structural diversities of bacteria are essential in understanding the soil biodiversity, hence forming one of the indicators of soil biodiversity. Bacterial communities are found to be the most reliable indicators of soil biodiversity and, thus, soil health. This is because the functionalities of the soil environment are influenced by soil bacterial communities both directly and indirectly [87]. Environmental factors are the dominant drivers in the composition of bacterial communities [88]. The environmental perturbations affect the chemical, physical, and biological activities of soil, which in turn affects the composition of the bacterial communities.

The changes in soil moisture, temperature, pH, C:N ratio, P, *etc.* alter the bacterial communities. Hence, different stresses on the soil environment are reliably portrayed by the bacterial community [89]. Environmental changes result in changes in the functional diversity of bacterial communities. Nutrients are essential resources for the soil's bacterial community. The structural diversity of bacteria can be studied by the ratio of oligotrophs to copiotrophs in soil. The data can be used as an indicator of the nutrient stress reflectance of the bacterial species in the soil [90]. The higher the ratio, the better the stability of environmental conditions will be and *vice versa.* A lower ratio is obtained in the case of soils contaminated with pesticides or sewage sludge.

Indicators of Carbon Cycling

Carbon cycling is an important soil ecosystem parameter, which is indicated by soil respiration, decomposition of organic matter, soil enzymes, methane oxidation to understand soil ecosystem health, microbial community health, and atmospheric balance.

Soil Respiration

Soil respiration indicates microbial activity in the soil, which includes soil bacterial activity too, as soil respiration is correlated with the carbon cycle. Thus, soil respiration has a clear relationship between soil organic matter content, its decomposition, microbial biomass, and activity. Soil respiration gives a clear indication of the contamination of soils either by pesticides or heavy metals [91].

Organic Matter Decomposition

Microbial activity results in organic matter decomposition. Hence, any change in microbial activity will result in changes in organic matter decomposition and nutrient cycling such as N, S, and P. Cellulose present in organic matter is degraded by the cellulase enzyme, which is produced by bacteria such as *Pseudomonas fluorescens*, *Bacillus subtilis*, *E. coli*, and *Serratia marcescens* [92]. Lignins in organic matter are mainly degraded by fungi. However, some actinomycetes, α and γ proteobacteria are found to degrade lignin due to the use of extracellular enzymes and secondary metabolites [93].

Soil Enzymes

Enzymes play an important role in essential biogeochemical processes in the soil, such as decomposition, nutrient cycling, immobilization, and mineralization. Soil enzymatic activity thus presents a useful indicator of soil microbial activity and soil health. Enzymes are involved in the decomposition of organic matter, the oxidation of methane, N cycling, *etc.*, which are very important in understanding soil health.

Methane Oxidation

Methane is produced by methanogenic bacteria and consumed by methanotrophic bacteria [94]. Methane production is linked to the atmospheric balance due to its status as a greenhouse gas, contributing to climate change. Hence, methane oxidation and methanotrophic bacteria serve as important indicators of soil as well as atmospheric quality. These bacteria consume methane and thus mitigate its effects as a potent greenhouse gas. This process is also carried out by enzymes such as methane monooxygenase, methanol-, formaldehyde-, and formate dehydrogenases [95].

Indicators of Nitrogen Cycling

Nitrogen cycling is another parameter of soil ecosystem health which is indicated by processes like N-mineralisation, nitrification, denitrification, and N fixation, which are essential in maintaining soil ecosystem and plant health, atmospheric balance, understanding leaching to groundwater and surface run-off. Soils rich in nitrifying bacteria will have a reasonable amount of nitrates in solution. Ammonification and nitrification are two major processes occurring in such soils, whereas anaerobic soils are proficient in denitrifying bacteria. Denitrification gives an idea of microbial biomass. Denitrification is the process of producing nitrogen gas from inorganic nitrogen compounds, typically a reduction process. It is affected by different soil management practices, thus it is important to understand soil health, emphasizing its role in soil ecosystem functions. Bacteria of the genus *Pseudomonas*, *Bacillus*, and *Alcaligenes* are the main denitrifiers. Symbiotic N fixers can also be considered indicators of soil health. The number of *Rhizobium* can be considered as an indicator of soil health, mainly to understand the nature of heavy metal contamination in agricultural soils [96]. Cyanobacteria (blue-green algae) carry out non-symbiotic N fixation, and the number of cyanobacteria is also considered as an indicator of soil health, particularly in soils contaminated by heavy metals. However, it is not considered a very stable indicator due to its high sensitivity to environmental conditions [91].

Indicators of Soil Biomass

Soil biomass provides an idea of soil health, which is understood by microbial biomass and protozoal biomass. Bacterial biomass is one of the parameters of microbial biomass, which contributes to soil biomass. Bacterial biomass, decomposition, and N mineralization are interrelated [97]. The soil biomass influences soil structure and soil stabilization [98]. The soil biomass is an indicator of soil carbon, which is a very potential indicator of soil health and soil quality. The ratio of fungal to bacterial biomass is another suitable indicator for understanding soil health for soil management studies.

Indicators of Microbial Activity

Microbial activity is used to understand the measurements at the soil ecosystem level by understanding their roles in nutrient cycles and the decomposition of organic matter. This is carried out by bacterial DNA and protein synthesis at the community level and by the activity of bacteriophages at the population level. Bacterial cells contain more protein than DNA, hence protein synthesis is found to be a more reliable indicator of bacterial activity [99]. RNA plays a major role in protein synthesis. This explains that the amount of RNA is an indicator of protein synthesis and, hence, bacterial activity. Bacteriophages are viruses that infect and multiply in a specific host bacterium. The bacteriophages depend on the activity of soil bacteria. Hence measuring the frequency of bacteriophages helps to understand the activity of soil bacteria.

Indicators of Key Species

Key species in soil form a soil ecosystem parameter to understand soil, plant, animal, and human health. Human pathogenic bacteria in soil are one important group, which determines the chances of human infection and serves as an indicator of human health. The presence of such pathogenic bacteria is thus an indicator of soil health. The presence of *E. coli* in the soil is an example of an indicator of fecal contamination, indicating deterioration in soil health. This could also serve as an indicator of the presence of other harmful bacteria in the soil. Enumeration of such pathogenic bacteria is thus essential in assessing soil health and its link to human health.

Indicators of Bioavailability

The bioavailability of chemical compounds in the soil is an ecosystem parameter, which can be measured by microbial indicators at community and population levels, like biosensor bacteria, the plasmid containing bacteria, antibiotic-resistant bacteria, *etc*. Biosensor bacteria are those that give a specific response to stress through their reporter genes. These are utilized in monitoring ecotoxicity in the soil, mainly through heavy metals, thus proving themselves as good indicators of soil health. The plasmid containing bacteria serve as one of the best indicators of environmental contamination as polluted soils possess a higher frequency of these bacteria [100]. The presence of antibiotic-resistant bacteria in soil is yet another indicator of poor soil health, indicating the risk to animal and human health, and urban and industrial pollution on the chances of surface run-off.

IMPACT OF AGRICULTURAL PRACTICES ON SOIL MICROBIAL HABITAT

A perusal of the contents of this chapter brings to light the importance of microbes, especially bacteria, in building soil health. They are our best known allies for a nutrient-rich forte for the plant kingdom. They not only play a significant role in nutrient cycling but are also good candidates for disease suppression and soil aggregate formation. Moreover, they can be used as indicators for analyzing soil health. The ability of these microorganisms to survive even in the extremes of the environment highlights their paramount importance in soil management. Therefore, we must preserve this natural microflora. Of late, there has been a noticeable change in the structure, chemistry, and biological activity of soils with the continuous use of land for crop production as well as for industrial purposes. The pressure is likely to continue owing to the rapidly increasing population. The challenge we face today is how well we can maintain soil health amidst the ever-increasing use of soil as a resource for various purposes.

The major pressures brought to bear on the composition of the soil microbial community are cultivation, fertilizers, pesticide residues, *etc*. Traditional cultivation methods, which include deep ploughing, lead to the formation of more compact soil. The intensive mixing also leads to the loss of organic matter and hastens humus mineralization, resulting in an altered microflora. Several years of research conducted by Gajd and Przewtkoka [101] revealed a decrease in the amount of microbial biomass carbon in conventional soil cultivation with a steady level of biomass in reduced tillage and direct sowing. Overuse of inorganic fertilizers and pesticides has also been shown to disturb the normal soil microbial habitat. Agricultural practices that use high amounts of inorganic fertilizers, pesticides, and other amendments can overcome specific soil constraints and can lead to an overall increase in food production. But it can also lead to deterioration in soil quality due to compaction, erosion, nutrient depletion, *etc*. Continuous use of chemicals in the soil can lead to loss of biodiversity and beneficial crop-associated biodiversity, resulting in pest and disease outbreaks.

CONCLUDING REMARKS

Agricultural sustainability is a measure of soil health. Healthy soil should provide a perspective of an eco-friendly environment and maximum crop yield potential even under changing management practices or natural calamities. Soil health can be enhanced by management and land-use decisions that consider the multiple functions of soil and that take into account that soil is a living organism. The way forward for us is sustainable agriculture. Minimal or zero tillage farming and organic agricultural practices can help recover biodiversity over a period of time. Switching from inorganic fertilizers to animal manures, sewage sludges, and composts has the potential not only to increase soil organic matter stocks but also to increase the soil microbial diversity. With regards to this context, biofertilizers, especially plant growth-promoting rhizobacteria, play a pivotal role in restoring soil health. PGPR not only enhances plant growth but also helps plants cope with biotic and abiotic stresses. Examples of these include seed and root inoculation with beneficial rhizobacteria such as *Azotobacter*, *Rhizobium*, *etc*. for enhanced soil fertility and inoculation of the soil and environment with biocontrol agents such as *Pseudomonas fluorescens*, *Bacillus subtilis*, *etc*. The addition of organic manure, compost, green manuring, and cropping system management are some indirect methods of modifying the microbial habitat.

Soil fertility is dependent upon three interacting and mutually dependent components: physical fertility, chemical fertility, and biological fertility, of which the latter is the most understudied topic. Microorganisms provide life for the soil. A million beneficial bacteria in 1 gram of soil is all you need to keep the soil healthy. Therefore, there is a need to protect, preserve, and conserve this natural niche for better yields and environmental quality through a new paradigm of crop production.

CONSENT FOR PUBLICATION

Not applicable.

CONFLICT OF INTEREST

The author declares no conflict of interest, financial or otherwise.

ACKNOWLEDGEMENTS

Declared none.

REFERENCES

[1]   Aislabie J, Deslippe JR. Soil Microbes and Their Contribution to Soil Services. In: Dymond JR, Ed. Ecosystem Services in New Zealand – Conditions and Trends. Lincoln, New Zealand: Manaaki Whenua Press 2013; pp. 143-61.

[2]   Fierer N, Schimel JP, Holden PA. Variation in microbial community composition through two soil depth profile. Soil Biol Biochem 2003; 35: 167-76.

[3]   Stumm W, Morgan JJ, Eds. Aquatic Chemistry Chemical Equilibria and Rates in Natural Waters. 3rd ed. NJ, USA: John Wiley & Sons, Inc 1996; p. 1040.

[4]   Maestrojuàn GM, Boone DR. Characterization of *Methanosarcina barkeri* MS^T and 227, *Methanosarcina mazei* S-6^T, and *Methanosarcina vacuolata* Z-761^T. Int J Syst Evol Microbiol 1991; 41: 267-74.

[5]   Linn DM, Doran JW. Effect of water-filled pore space on carbon dioxide and nitrous oxide production in tilled and nontilled soils. Soil Sci Soc Am J 1984; 48: 1267-72.

[6]   Santana MM, Gonzalez JM, Clara MI. Inferring pathways leading to organic-sulfur mineralization in the Bacillales. Crit Rev Microbiol 2016; 42: 31-45.
        [PMID: 24506486]

[7]   Holden JF. Extremophiles: Hot Environments. In: Schaechter M, Ed. Encyclopedia of Microbiology. 3rd ed. Oxford, United Kingdom: Academic Press, Elsevier 2009; pp. 127-46.

[8]   Margesin R, Miteva V. Diversity and ecology of psychrophilic microorganisms. Res Microbiol 2011; 162: 346-61.
        [PMID: 21187146]

[9]   Zhou MY, Zhang YJ, Zhang XY, *et al. Flavobacterium phocarum* sp. nov., isolated from soils of a seal habitat in Antarctica. Int J Syst Evol Microbiol 2018; 68: 536-41.
        [PMID: 29251588]

[10]   Tyagi S, Singh DK. *Azospirillum himalayense* sp. nov. A nifH bacterium isolated from Himalayan valley soil, India. Ann Microbiol 2014; 64: 259-66.

[11] Johnson DB. Microorganisms and the Biogeochemical Cycling of Metals in Aquatic Environments. In: Langston WJ, Bebianno MJ, Eds. Metal Metabolism in Aquatic Environments. Boston, MA: Springer 1998; pp. 31-57.

[12] Nielsen MN, Winding A. Microorganisms as indicators of soil health. National Environmental Research Institute Denmark. Technical Report No. 388 2002.

[13] Niu S, Classen AT, Luo Y. Functional traits along a transect. Funct Ecol 2018; 32: 4-9.

[14] Torsvik V, Øvreås L, Thingstad TF. Prokaryotic diversity--magnitude, dynamics, and controlling factors. Science 2002; 296: 1064-6.
[PMID: 12004116]

[15] Winogradsky S. Sur la microflora autochtone de la terre arable. Comptes rendus hebdomadaires des seances de l'Academie des Sciences (Paris) D 1924; 178: 1236-39.

[16] Van Gestel M, Merckx R, Vlassak K. Microbial biomass responses to soil drying and rewetting: the fate of fast- and slow growing microorganisms in soils from different climates. Soil Biol Biochem 1993; 25: 109-23.

[17] Odum EP. The strategy of ecosystem development. Science 1969; 164: 262-70.
[PMID: 5776636]

[18] Pianka ER. On r- and K-selection. Am Nat 1970; 104: 592-7.

[19] Van Elsas JD, Jansson JK, Trevors JT, Eds. Modern Soil Microbiology. Boca Raton, FL, USA: CRC Press 2007; 2: pp. 237-61.

[20] Liles MR, Manske BF, Bintrim SB, Handelsman J, Goodman RM. A census of rRNA genes and linked genomic sequences within a soil metagenomic library. Appl Environ Microbiol 2003; 69: 2684-91.
[PMID: 12732537]

[21] Janssen PH. Identifying the dominant soil bacterial taxa in libraries of 16S rRNA and 16S rRNA genes. Appl Environ Microbiol 2006; 72: 1719-28.
[PMID: 16517615]

[22] Goldfarb KC, Karaoz U, Hanson CA, *et al.* Differential growth responses of soil bacterial taxa to carbon substrates of varying chemical recalcitrance. Front Microbiol 2011; 2: 94.
[http://dx.doi.org/10.3389/fmicb.2011.00094] [PMID: 21833332]

[23] Ward NL, Challacombe JF, Janssen PH, *et al.* Three genomes from the phylum Acidobacteria provide insight into the lifestyles of these microorganisms in soils. Appl Environ Microbiol 2009; 75: 2046-56.
[PMID: 19201974]

[24] Hill P, Krištůfek V, Dijkhuizen L, Boddy C, Kroetsch D, van Elsas JD. Land use intensity controls actinobacterial community structure. Microb Ecol 2011; 61: 286-302.
[PMID: 20924760]

[25] Kulichevskaia IS, Belova SE, Kevbrin VV, *et al.* Analysis of the bacterial community developing in the course of *Sphagnum* moss decomposition. Mikrobiologiia 2007; 76: 702-10.
[PMID: 18069332]

[26] Humbert S, Tarnawski S, Fromin N, *et al.* Molecular detection of anammox bacteria in terrestrial ecosystems: distribution and diversity. ISME J 2010; 4: 450-4.
[PMID: 20010634]

[27] Fuerst JA, Sagulenko E. Beyond the bacterium: planctomycetes challenge our concepts of microbial structure and function. Nat Rev Microbiol 2011; 9: 403-13.
[PMID: 21572457]

[28]   Tian JH, Pourcher AM, Bouchez T, *et al.* Occurrence of lignin degradation genotypes and phenotypes among prokaryotes. Appl Microbiol Biotechnol 2014; 98: 9527-44.
[PMID: 25343973]

[29]   Pajares S, Bohannan BJM. Ecology of nitrogen fixing, nitrifying, and denitrifying microorganisms in tropical forest soils. Front Microbiol 2016; 7: 1045.
[http://dx.doi.org/10.3389/fmicb.2016.01045] [PMID: 27468277]

[30]   Babalola OO, Glick BR. The use of microbial inoculants in African agriculture: current practice and future prospects. J Food Agric Environ 2012; 10: 540-9.

[31]   Glick BR. Plant growth-promoting bacteria: mechanisms and applications. Scientifica (Cairo) 2012; 2012: 963401.
[http://dx.doi.org/10.6064/2012/963401] [PMID: 24278762]

[32]   Nadeem SM, Zahir ZA, Naveed M, *et al.* Preliminary investigations on inducing salt tolerance in maize through inoculation with rhizobacteria containing ACC deaminase activity. Can J Microbiol 2007; 53(10): 1141-9.
[PMID: 18026206]

[33]   Sziderics AH, Rasche F, Trognitz F, *et al.* Bacterial endophytes contribute to abiotic stress adaptation in pepper plants (*Capsicum annuum* L.). Can J Microbiol 2007; 53(11): 1195-202.
[PMID: 18026213]

[34]   Madhaiyan M, Poonguzhali S, Sa T. Metal tolerating methylotrophic bacteria reduces nickel and cadmium toxicity and promotes plant growth of tomato (*Lycopersicon esculentum* L). Chemosphere 2007; 69: 220-8.
[PMID: 17512031]

[35]   Moyne AL, Shelby R, Cleveland TE, *et al.* Bacillomycin D: an iturin with antifungal activity against *Aspergillus flavus.* J Appl Microbiol 2001; 90: 622-9.
[PMID: 11309075]

[36]   Raaijmakers JM, Weller DM. Natural plant protection by 2, 4- diacetylphloroglucinol-producing *Pseudomonas* spp. in take-all decline soils. Mol Plant Microbe Interact 1998; 11: 144-52.

[37]   Reinhold BB, Chan SY, Reuber TL, *et al.* Detailed structural characterization of succinoglycan, the major exopolysaccharide of *Rhizobium meliloti* Rm1021. J Bacteriol 1994; 176: 1997-2002.
[PMID: 8144468]

[38]   Holban AM, Grumezescu AM, Eds. Microbial Production of Food Ingredients and Additives. Academic Press 2017; p. 518.

[39]   Arnon DI, Stout PR. The essentiality of certain elements in minute quantity for plants with special reference to copper. Plant Physiol 1939; 14: 371-5.
[PMID: 16653564]

[40]   Hansell DA, Carlson CA. Marine dissolved organic matter and the carbon cycle. Oceanography (Wash DC) 2001; 14: 41-9.

[41]   Li Y, Wu J, Shen J, *et al.* Soil microbial C:N ratio is a robust indicator of soil productivity for paddy fields. Sci Rep 2016; 6: 35266.
[http://dx.doi.org/10.1038/srep35266] [PMID: 27739462]

[42]   Gruber N, Galloway JN. An Earth-system perspective of the global nitrogen cycle. Nature 2008; 451: 293-6.
[PMID: 18202647]

[43]   Karthika KS, Rashmi I, Parvathi MS. Biological functions, uptake and transport of essential nutrients in relation to plant growth. Hasanuzzaman M, Fujita M, Oku H, *et al.* Eds. Plant Nutrients and Abiotic Stress Tolerance.. Singapore: Springer 2018; pp. 1-49.

[44]   Smercina DN, Evans SE, Friesen ML, *et al.* To fix or not to fix: controls on free-living nitrogen

fixation in the rhizosphere. Appl Environ Microbiol 2019; 85: e02546-18.
[http://dx.doi.org/10.1128/AEM.02546-18] [PMID: 30658971]

[45]   Sulieman S, Tran LS. Symbiotic nitrogen fixation in legume nodules: metabolism and regulatory mechanisms. Int J Mol Sci 2014; 15: 19389-93.
[PMID: 25347276]

[46]   James EJ. Nitrogen fixation in endophytic and associative symbiosis. Field Crops Res 2000; 65: 197-209.

[47]   White JF Jr, Crawford H, Torres MS, *et al.* A proposed mechanism for nitrogen acquisition by grass seedlings through oxidation of symbiotic bacteria. Symbiosis 2012; 57: 161-71.
[PMID: 23087539]

[48]   Indian Society of Soil Science, Eds. Soil Science: An introduction. New Delhi: Indian Society of Soil Science 2015; p. 728.

[49]   Luo J, Tillman RW, Ball PR. Nitrogen loss through denitrification in a soil under pasture in New Zealand. Soil Biol Biochem 2000; 32: 497-509.

[50]   Strong DT, Fillery IRP. Denitrification response to nitrate concentrations in sandy soils. Soil Biol Biochem 2002; 34: 945-54.

[51]   Richardson AE, Simpson RJ. Soil microorganisms mediating phosphorus availability update on microbial phosphorus. Plant Physiol 2011; 156: 989-96.
[PMID: 21606316]

[52]   Oliveira CA, Sà NMH, Gomes EA, *et al.* Assessment of the mycorrhizal community in the rhizosphere of maize (*Zea mays* L.) genotypes contrasting for phosphorus efficiency in the acid savannas of Brazil using denaturing gradient gel electrophoresis (DGGE). Appl Soil Ecol 2009; 41: 249-58.

[53]   Klotz MG, Bryant DA, Hanson TE. The microbial sulfur cycle. Front Microbiol 2011; 2: 241.
[http://dx.doi.org/10.3389/fmicb.2011.00241] [PMID: 22144979]

[54]   Cachada A, Rocha-Santos T, Duarte AC. Soil and Pollution: An Introduction to the Main Issues. In: Duarte AC, Cachada A, Rocha-Santos T, Eds Soil Pollution- From Monitoring to Remediation Acdemic Press, London, UK. 2018; pp. 1-28.

[55]   FAO and ITPS. Status of the World's Soil Resources (SWSR) - Main Report. Rome, Italy: Food and Agriculture Organization of the United Nations and Intergovernmental Technical Panel on Soils 2015.

[56]   Gupta R, Mohapatra H. Microbial biomass: an economical alternative for removal of heavy metals from waste water. Indian J Exp Biol 2003; 41: 945-66.
[PMID: 15242288]

[57]   Barr D, Finnamore JR, Bardos RP, *et al.* Eds. Biological Methods for Assessment and Remediation of Contaminated Land: Case Studies.. Westminster, London: CIRIA 2002; p. 178.

[58]   Atlas RM, Hazen T, Philp JC, Eds. Bioremediation Applied microbial solutions for real-world environmental cleanup. ASM Press 2005; p. 370.

[59]   Akthar MN, Sastry KS, Mohan PM. Mechanism of metal ion biosorption by fungal biomass. Biometals 2007; 9: 21-8.

[60]   Kisielowska E, Hołda A, Niedoba T. Removal of heavy metals from coal medium with application of biotechnological methods. Górnictwo i Geoinżynieria 2010; 34: 93-104.

[61]   Zhang H, Ma D, Qiu R, *et al.* Non-thermal plasma technology for organic contaminated soil remediation: a review. Chem Eng J 2017; 313: 157-70.

[62]   Chen S, Hu M, Liu J, *et al.* Biodegradation of beta-cypermethrin and 3-phenoxybenzoic acid by a novel Ochrobactrum lupini DG-S-01. J Hazard Mater 2011; 187: 433-40.
[PMID: 21282001]

[63]   Zhang C, Jia L, Wang S, *et al.* Biodegradation of beta-cypermethrin by two *Serratia* spp. with

different cell surface hydrophobicity. Bioresour Technol 2010; 101: 3423-9.
[PMID: 20116237]

[64]   Ghosal D, Ghosh S, Dutta TK, *et al.* Current state of knowledge in microbial degradation of polycyclic aromatic hydrocarbons (PAHs): a review. Front Microbiol 2016; 7: 1369.
[http://dx.doi.org/10.3389/fmicb.2016.01369] [PMID: 27630626]

[65]   Morehead NR, Eadie BJ, Lake B, *et al.* The sorption of PAH onto dissolved organic matter in Lake Michigan waters. Chemosphere 1986; 15: 403-12.

[66]   Foght J. Anaerobic biodegradation of aromatic hydrocarbons: pathways and prospects. J Mol Microbiol Biotechnol 2008; 15: 93-120.
[PMID: 18685265]

[67]   Mallick S, Chakraborty J, Dutta TK. Role of oxygenases in guiding diverse metabolic pathways in the bacterial degradation of low-molecular-weight polycyclic aromatic hydrocarbons: a review. Crit Rev Microbiol 2011; 37: 64-90.
[PMID: 20846026]

[68]   Baker K, Cook RJ, Eds. Biological Control of Plant Pathogens. SF, USA: WH Freeman and Company 1974; p. 433.

[69]   Mendes R, Kruijt M, de Bruijn I, *et al.* Deciphering the rhizosphere microbiome for disease-suppressive bacteria. Science 2011; 332: 1097-100.
[PMID: 21551032]

[70]   Yadav RS, Panwar J, Meena HN, *et al.* Developing Disease-Suppressive Soil Through Agronomic Management. In: Meghvansi M, Varma A, Eds. Organic Amendments and Soil Suppressiveness in Plant Disease Management. Cham: Springer 2015; pp. 61-94.

[71]   Ulloa-Ogaz AL, Muñoz-Castellanos LN, Nevárez-Moorillón GV. Biocontrol of Phytopathogens: Antibiotic Production as Mechanism of Control. In: Méndez-Vilas A, Ed. The Battle Against Microbial Pathogens: Basic Science. Technological Advances and Educational Programs 2015; pp. 305-9.

[72]   Latz E, Eisenhauer N, Rall BC, *et al.* Plant diversity improves protection against soil-borne pathogens by fostering antagonistic bacterial communities. J Ecol 2012; 100: 597-604.

[73]   Malajczuk N, Theodorou C. Influence of water potential on growth and cultural characteristics of *Phytophthora cinnamomi*. Trans Br Mycol Soc 1979; 72: 15-8.

[74]   McNeely D, Chanyi RM, Dooley JS, *et al.* Biocontrol of *Burkholderia cepacia* complex bacteria and bacterial phytopathogens by *Bdellovibrio bacteriovorus*. Can J Microbiol 2017; 63: 350-8.
[PMID: 28177793]

[75]   van Loon LC, Bakker PAHM, Pieterse CMJ. Systemic resistance induced by rhizosphere bacteria. Annu Rev Phytopathol 1998; 36: 453-83.
[PMID: 15012509]

[76]   Brahmaprakash GP, Sahu PK, Lavanya G, *et al.* Microbial Functions of theRhizosphere. In: Singh D, Singh H, Prabha R, Eds. Plant-Microbe Interactions In Agro-Ecological Perspectives. Singapore: Springer 2017; pp. 177-210.

[77]   Garnica-Vergara A, Barrera-Ortiz S, Muñoz-Parra E, *et al.* The volatile 6-pentyl-2H-pyran-2-one from *Trichoderma atroviride* regulates *Arabidopsis thaliana* root morphogenesis *via* auxin signaling and ethylene insensitive 2 functioning. New Phytol 2016; 209: 1496-512.
[PMID: 26568541]

[78]   Choudhary DK, Prakash A, Johri BN. Induced systemic resistance (ISR) in plants: mechanism of action. Indian J Microbiol 2007; 47: 289-97.
[PMID: 23100680]

[79]   Six J, Bossuyt H, Degryze S, *et al.* A history of research on the link between (micro) aggregates, soil

biota, and soil organic matter dynamics. Soil Tillage Res 2004; 79: 7-31.

[80]   Rillig MC, Muller LA, Lehmann A. Soil aggregates as massively concurrent evolutionary incubators. ISME J 2017; 11: 1943-8.
[PMID: 28409772]

[81]   Wilpiszeski RL, Aufrecht JA, Retterer ST, *et al.* Soil aggregate microbial communities: towards understanding microbiome interactions at biologically relevant scales. Appl Environ Microbiol 2019; 85: e00324-19.
[http://dx.doi.org/10.1128/AEM.00324-19] [PMID: 31076430]

[82]   Nishio M, Furusaka C. The distribution of nitrifying bacteria in soil aggregates. Soil Sci Plant Nutr 1970; 16: 24-9.

[83]   Wingender J, Neu TR, Flemming HC. What are Bacterial Extracellular Polymeric Substances? In: Wingender J, Neu TR, Flemming HC, Eds. Microbial Extracellular Polymeric Substances. Berlin, Heidelberg: Springer 1999; pp. 1-19.

[84]   Costa OYA, Raaijmakers JM, Kuramae EE. Microbial extracellular polymeric substances: ecological function and impact on soil aggregation. Front Microbiol 2018; 9: 1636.
[PMID: 30083145]

[85]   Shida T, Mukaijo K, Ishikawa S, *et al.* Production of long-chain levan by a sacC insertional mutant from *Bacillus subtilis* 327UH. Biosci Biotechnol Biochem 2002; 66: 1555-8.
[PMID: 12224641]

[86]   Kumar KSA, Karthika KS. Abiotic and Biotic Factors Influencing Soil Health and/or Soil Degradation. In: Giri B, Varma A, Eds. Soil Health Soil Biology. Cham: Springer 2020; Vol. 59: pp. 145-61.

[87]   Wu T, Chellemi DO, Graham JH, *et al.* Comparison of soil bacterial communities under diverse agricultural land management and crop production practices. Microb Ecol 2008; 55: 293-310.
[PMID: 17619214]

[88]   Lear G, Washington V, Neale M, *et al.* The biogeography of stream bacteria. Glob Ecol Biogeogr 2013; 22: 544-54.

[89]   Hermans SM, Buckley HL, Case BS, *et al.* Bacteria as emerging indicators of soil condition. Appl Environ Microbiol 2016; 83: e02826-16.
[http://dx.doi.org/10.1128/AEM.02826-16] [PMID: 27793827]

[90]   van Brüggen AHC, Semenov AM. In search of biological indicators for soil health and disease suppression. Appl Soil Ecol 2000; 15: 13-24.

[91]   Brookes PC. The use of microbial parameters in monitoring soil pollution by heavy metal. Biol Fertil Soils 1995; 19: 269-79.

[92]   Sethi S, Datta A, Gupta BL, *et al.* Optimization of cellulase production from bacteria isolated from soil. ISRN Biotechnol 2013; 2013: 985685.
[http://dx.doi.org/10.5402/2013/985685] [PMID: 25937986]

[93]   Huang XF, Santhanam N, Badri DV, *et al.* Isolation and characterization of lignin-degrading bacteria from rainforest soils. Biotechnol Bioeng 2013; 110: 1616-26.
[PMID: 23297115]

[94]   Ritchie DA, Edwards C, McDonald IR, *et al.* Detection of methanogens and methanotrophs in natural environments. Glob Change Biol 1997; 3: 339-50.

[95]   Hooper AB. Chemolithotrophy. Encyclopedia of Biological Chemistry. Academic Press, Elsevier. 2013; pp. 419-24.

[96]   Nielsen MN, Winding A. Microorganisms as Indicators of Soil Health. National Environmental Research Institute, Denmark 2002; 388: 84.

[97]   Carter MR, Gregorich EG, Angers DA, *et al.* Interpretation of microbial biomass measurements for soil quality assessment in humid temperate regions. Can J Soil Sci 1999; 79: 507-20.

[98]   Alföldi T, Lockeretz W, Niggli U, Eds. IFOAM 2000, The World Grows Organic: Proceedings 13th International IFOAM Scientific Conference, IFOAM, Hochschulverlag, Zürich, Switzerland. 2000; p. 762.

[99]   Bååth E. Growth rates of bacterial communities in soils at varying pH: a comparison of the thymidine and leucine incorporation technique. Microb Ecol 1998; 36: 316-27.
[PMID: 9852511]

[100]   Drønen AK, Torsvik V, Goksøyr J, *et al.* Effect of mercury addition on plasmid incidence and gene mobilizing capacity in bulk soil. FEMS Microbiol Ecol 1998; 27: 381-94.

[101]   Gajda A, Przewloka B. Soil biological activity as affected by tillage intensity. Int Agrophys 2012; 26: 15-23.

CHAPTER 2

An Industrial Diligence of Behooveful Soil Microorganisms

Someshree S. Mane[1,*], Pravin B. Khaire[1] and **Sandesh V. Pawar[2]**

[1] *Department of Plant Pathology and Agril. Microbiology, PGI, MPKV, Rahuri, India*

[2] *Department of Plant Pathology, Sumitomo Chemical India Limited, Mumbai, India*

Abstract: The increasing need for environmentally friendly products or substances is driving the use of metabolites based on beneficial microorganisms. Soil is essential for the maintenance of biodiversity above and below ground. Millions of microorganisms live and reproduce in a few grams of topsoil, an ecosystem essential for life on earth. Moreover, microorganisms are capable of producing chemical compounds that have not been synthesized in the laboratory or can only be processed with considerable difficulty. Not only do these soil microflorae play a significant role in the conservation of soil quality, but they also play a vital role in biomedical, pharmaceutical, and industrial applications. In this chapter, we address recent advances in the industrial manufacture of chemical products by microorganisms.

Keywords: Industrial use, Natural products, Soil microbes.

INTRODUCTION

Microorganisms first appeared on earth a billion years ago and have now developed and increased enormously into new ecosystems globally. Their existence has driven new ecosystems to grow, some of which have allowed more complex species to evolve. Microorganisms are particularly important in industrial microbiology to synthesize a variety of items that are beneficial to humans and have significantly altered our lives and life spans. Beverages, dietary additives, human and animal health products, and biofuels are included in these products. Moreover, microorganisms are capable of producing chemical compounds that have not been synthesized in the laboratory or can only be processed with considerable difficulty. Consequently, the action of microorganisms on inexpensive carbohydrate materials may render certain chemical products cheaper than those of other chemical compounds. In this chap-

---

* **Corresponding author Someshree S. Mane:** Department of Plant Pathology and Agril. Microbiology, PGI, MPKV, Rahuri, India; E-mail: manesomeshree70@gmail.com

ter, we address recent advances in the industrial manufacture of chemical products by microorganisms.

Recovery of pivotal nutrients on earth will cease without metabolism and interaction between microorganisms. When subjected to severe conditions, microorganisms adjust and evolve, producing a variety of secondary metabolites that primarily mediate important functions in generating microorganisms' life. Secondary metabolites (SMs) have been found to have disease control potential in surgical, biomedical, and industrial applications. The mechanism by which a living microorganism modifies or attaches simple molecules to form macromolecules is known as biosynthetic pathways [1]. Natural compounds are unique, organic compounds and naturally existing substances produced by microorganisms or byproducts of secondary metabolism.

Natural ingredients have been used in the preparation of medicines, cosmetics, dietary supplements, and raw materials [2]. This section covers many biological processes and SMs produced by microorganisms. The future relevance of natural products is often explored in biomedical, medical, and industrial biotechnologies.

Soil is an ecosystem capable of producing the resources needed for living organisms to develop. Soil microorganisms (bacteria and fungi) are responsible for the decomposition of biomass, the circulation of biogenic elements that make nutrients available to plants, impurity biodegradation, and soil structure maintenance. The presence of soil microorganisms depends on their chemical composition, moisture, pH, and structure. Microbes are being used for the development of vitamins, amino acids, and enzymes. Microorganisms are used to make pharmaceuticals that may not be produced otherwise.

METABOLISM OF MICROBES

Bacterial metabolic substances and compounds are microbial metabolites. Bacteria have been widely used in biological science owing to their increasing, simple proliferation, and comparatively straightforward cellular structure. Bacterial metabolism has been researched much more to determine how to prevent or interrupt their development and toxic metabolites. Microbial metabolic products have been classified as primary metabolites (PMs) and secondary metabolites (SMs) [3]. Microbiological biosynthesis is divided into dietary classes based on the following three key factors [3]:

-The energy source used for growth.
-The carbon source.
-The electron donor source used for growth.

Primary Metabolites (PMs)

Substances formed during a culture's exponential phase are known as PMs [4]. A primary metabolite is a critical component in maintaining normal physiological processes, and hence, it is often referred to as a core metabolite [5]. PMs are produced as byproducts of the exponential process and are involved in formation, evolution, and regeneration, making them essential for their sustainability and life. PMs are known as core metabolites because they sustain regular physiological processes as well [4]. The developmental pattern is consistent with the growth path in primary metabolism, and metabolites such as supplements, proteins, and nucleosides are produced in sufficient amounts to support cell growth [5]. The PMs required for cell creation are proteins, nucleic acids, fats, lipoproteins, and supplements, while ethanol, acetone-butanol, and organic acids are required for power generation. Citric acid is a common example of a PM used in industrial microbiology.

Secondary Metabolites (SMs)

SMs are produced and marked the start of the steady-state growth process; they do not participate in tissue formation, evolution, or multiplication [5]. Most of the recognized SMs play a role in natural ecosystems, such as functioning as defence mechanisms, antibiotics, or pigment producers. Whenever one or more nutrients in growth media are depleted, microbes emit secondary metabolites that usually have a crucial environmental function and a variety of survival functions. SMs are essential for human society's health, diet, and economy, even though they are not needed for microbial growth [6, 7]. SMs include toxins, gibberellins, alkaloids, antibiotics, and biopolymers.

Biocatalysis (BCTs)

BCTs, a major component of biotechnology, are essentially used to catalyze complex chemical processes. Microbial catalysts have long been used in the food sector, but modern technologies are emerging in a variety of fields, particularly analytical chemistry [8]. In BCTs entire microorganisms, cell extracts, refined enzymes, paralyzed cells, or paralyzed enzymes are used. Recent advances in large scale DNA sequencing, selective development, epigenetic modifications, synthetic biology, data processing, and biophysics have contributed to the rapid advancement of BCTs [9]. Cells manufacture complex protein molecules that function as BCTs. They possess catalytic properties that enable them to carry out complex chemical transitions, making them suitable for bioreactors in food, agricultural, and drug companies.

Biosynthesis

The biosynthesis approach is the synthesis of living organic compounds. During biosynthesis, basic substances are updated, transformed into other molecules, or blended into macromolecules. This system necessitates many metabolic processes, most of which contain enzymes found in cellular organelles [1]. Biosynthesis is triggered by a series of reactions that involve precursor compounds, catalytic enzymes, co-factors, and energy sources [1]. The interaction with the atmosphere and the specific chemical structures of biologically active microbial metabolites are their distinguishing features [10]. Natural compounds, including microbial metabolites, can be used practically in three ways: (1) directly contributing the fermentation product to medicine, agriculture, or some other field; (2) being used as a substrate for corresponding biochemical or microbial modification; and (3) microbial metabolites are being used as active molecules for the chemical synthesis of a variety of compounds [3]. The controlled biosynthetic processes in bacteria and fungi are associated with biomolecular interactions affecting various enzyme molecules as well as catalytic reactions and controlling various functions [11].

PRIMARY BIOSYNTHETIC PRODUCTS (PBPs)

PBPs are normal byproducts of exponential growth, where the production curve follows the trend line and metabolites (*e.g.* vitamins, amino acids, nucleosides) are developed to encourage cell growth [5]. The following are a few examples of PBPs:

Vitamins

Microbes use the chemical components of the culture medium to produce vitamins and other growth-inducing compounds. Such molecules are generated in abundance to meet their needs, processed in crops and then retrieved. Microorganisms are commercially processed to produce large-scale vitamins in a variety of cultural settings. Examples of microbiologically produced vitamins are carotene, the precursor of vitamin A from *Blakeslea trispora*, riboflavin from *Ashbya gossypii*, l-sarbose in vitamin C synthesis from *Gluconobacter oxydans*, and vitamin B12 from *Bacillus coagulans, Bacillus megaterium, Pseudomonas denitrificans, and Streptomyces olivaceus* [3, 10].

Amino Acids (AAs)

Glycine, threonine, glutamate, lysine, ornithine, and aspartate are AA's utilized by anaerobic bacteria, which can be metabolized to acetate [12]. Threonine, glutamate, and lysine are utilized to produce butyrate. Propionate was said to be primarily produced *via* threonine [13]. Just like the genetic manipulation of core metabolic processes, biosynthetic pathways and transport processes, it has been reported that metabolic processes and osmoregulations are gaining interest from researchers as new opportunities to be designed for the excess production of L-lysine. Additionally, species like *Corynebacterium glutamicum* and *Bacillus methanolicus* are gaining popularity in the manufacture of L-lysine utilizing methanol and lignocellulose as an important construction material [14].

Ethanol

Distilled ethanol possesses approximately 95% ethanol, which is normally sufficient for applications in the medicinal, cosmetic, agricultural, and chemical industries. Distillation cannot be a better option for ethanol concentration due to its liquid nature and vapour point with water. Even so, to be blended with fuel, ethanol must be at least 99.2% pure. That is achieved by dehydrating or soaking the ethanol after distillation [15]. Distillation can be accomplished by a variety of techniques, including solvent extraction, per-vaporation, liquid and vapour distillation, membrane technology, carbon dioxide extraction, and vapour recompression [16].

Acetone-Butanol

The commercial acetone-butanol fermentation process, which uses *Clostridium acetobutylicum*, was invented at Manchester University, UK in 1912 and it was easily expanded through the first and second world wars, due to its use in munitions manufacturing and later colour strain production [17, 18]. In the 1950s, the fermentation process fell out of favour in western developed countries because artificial substitutes could be generated at a cheaper cost *via* fuel sources. Few reported favorably, despite the fact that it persisted in China, Russia, and South Africa until the 1980s [18, 19].

Organic Acids

An environmental impetus to substitute conventional chemical methods with biologically determined organic acid processing has shown fungi as stunning cell factories. The natural synthesis of organic acids by fungi is believed to play a variety of important functions in nature based on the types of fungi that produce them.

Subsequently, various fungal species have been found and used in the processing of acids. Citric acid is the most popular transformational metabolic product. This acid is found in all animals and plants [20]. *Penicillium janthinellum* and the species of *Aspergillus*, *viz. A. foetidus*, *A. awamori*, *A. wentii*, and *A.*

carbonarius, can also form citric acid. It is a multipurpose and safe alimentary that is widely used in the pharmaceutical, cosmetic, and food industries due to its universal safety recognition, pleasant acid taste, high water solubility, and chelating properties [21].

SECONDARY BIOSYNTHETIC PRODUCTS

The fields of architecture, mechanical engineering, and natural systems are mostly eye-opening in terms of resolving technical challenges. Nature has created a variety of materials and natural products with incredible functional properties, mainly biopolymers. A natural substance is just generated by an ecosystem, which is not synthetic and the concept seems to be something nice [22].

There are two radically distinct forms of metabolic processes in living structures. The first is supposed to produce one or a few distinct molecules, both of which are pathways. The second is formed into a vast number of distinct metabolites [23]. Although the number of smaller molecules in living systems could be accountable to that of the 2nd class, which is called the diversity generation [23, 24]. There is no single end product to the variety of pathways; each enzyme in a diversity-generating pathway has a relaxed sustenance and can handle several compounds [25]. The following is discussed regarding the biosynthesis of certain key SMs produced by some microorganisms.

Biopolymers

Biopolymers are polymers that are tightly attached to large molecules by monomers. The prefix "bio" refers to degradable products produced by living things. The term biopolymer can be used for the identification of a vast variety of methods, typically generated from biological sources such as microorganisms, plants, *etc*. Green polymers are partly or entirely derived from natural materials that are sustainable [26]. Biopolymers are complex molecule assemblies that abide by specific and accurate 3D models and theories compared to synthetic polymers with an easier and more arbitrary structure. This is an important function, which enables biopolymers to work *in vivo* [27]. Exopolysaccharides are synthesized by multiple pathways of biosynthesis. In the genomes of their development forms, the genes responsible for synthesis are also listed [28]. Table **1** lists microbial polymers and their functions.

Table 1. Microbial polymers and their functional attributes [29].

Biopolymer	Bacterial Strain	Monomer Units	Applications
Alginate	*Pseudomonas aeruginosa,* and *Azotobacter vinelandii*	Guluronic acid and mannuronic acid	Food hydrocolloid and medicine: surgical dressings, hypoallergenic wound-healing tissue, and controlled drug release. As an immobilization matrix for viable cells and enzymes, and as the coating of the roots of seedlings and plants to prevent desiccation
Curdlan	*Rhizobium meliloti*, and *Agrobacterium radiobacter*	Glucose	Food, the pharmaceutical industry, heavy metal removal, and as a concrete additive, as a gelling agent, and immobilization matrix. It displays promising high antiretroviral activity (anti-AIDS drug)
Levan	*Bacillus, Streptococcus, Pseudomonas*, and *Zymomonas*	Fructose	For the production of sweet confectionery and ice cream as viscosifer and stabilizer

The use of biopolymers in the food industry as insulation, food packaging, and accepting matrices for healthy foods is now widely available for the industry. The growing interest in the use of bio-polymers in edible forms is attributed to their potential to provide a range of usable ingredients. The use of biopolymers increases the shelf life of the commodity and reduces the gross carbon footprint for food packaging [30]. Certain biomass capital materials, such as proteins, their ability as gas and scent barriers for polysaccharides and lipids (for instance, wax), are also excellent [29].

Biopolymers are also used to shield fragile compounds from the effects of harmful conditions during encapsulation. The term "microencapsulation" covers a given method of packing solids, liquids, or gases into small capsules that, under particular circumstances, can release their contents. The pharmaceutical industry has a great interest in such developments. However, the greatest benefit of coatings is that they can be used as a medium in which natural or chemically active materials, including antioxidants and antimicrobials, enzymes, or useful compounds, such as probiotics, minerals, and vitamins, may be added. The food will absorb these ingredients, which enhances protection, nutritional and sensory properties. Additionally, edible films can be used as flavour or scent carriers to provide a shield for aroma loss [31].

Bioflocculants

Colloids, cells, and suspended solids are applied to and eliminated from the suspension. Flocculation is a process. As a product of accumulation, the solids appear clearly like flowers or flakes. Actual or synthetic flocculants are compounds used in different manufacturing applications in the separation of solids and liquids by a flocculation device. For instance, the more particles are suspended, the higher the sedimentation speeds up. This would lead to an effective and simple method of flocculation [32]. A strong bioflocculant option would influence the high output of flocculation, the intensity of the aggregate particles, the number, and intensity of ties formed as a result of good flocculation [32]. Different industrial systems, such as potable water purification or wastewater recycling and downstream systems are typically employed in the fermentation field [33, 34]. In the fruit, beverage, and medicinal industries, microbial cells separate from bread may be used as a replacement for the filtration and centrifugation [32]. Bacteria in the genera *Pseudomonas* and *Bacillus* are significant sources of biosurfactants. However, the majority of biosurfactants of bacterial origin are ineffective for use in the food industry due to their potentially pathogenic nature. *Candida bombicola* and *C. lipolytica* are among the most highly studied yeasts for the development of biosurfactants.

Biofilm Synthesis

The tendency of microorganisms to make biofilms is an evolutionary survival mechanism that has occurred under extremely adverse environmental conditions. Biofilm arises by microorganisms as an evolutionary survival mechanism in response to extreme environmental changes, which can affect more than one animal. Biofilms require several regulatory networks, which convert signals into coordinated gene expression modifications to promote survival in adverse environments.

Another fascinating new technology is the use of the natural development of bacterial biofilms as a potential carrier, and not just for the development of inoculum, established bacterial or fungi-bacteria consortia. Biofilm processing is now possible for various industrial applications (*e.g.*, wastewater treatment, development of chemical compounds). There are four steps in the growth of a mature biofilm: initial attachment, permanent attachment to the processing of extracellular polysaccharides (EPs), early creation, and maturation of the biofilm architecture. The development of EP, which helps connect the cell to the surface and shield it from the environment, is highly important. EPs can be made up of polysaccharides, proteins, nucleic acids, or phospholipids. Exopolysaccharide alginate is a natural EPs formed by bacterial cells in biofilms [35].

Beneficial biofilms formed in *in vitro* cultures containing both fungal and bacterial strains have been used as biofertilizers for non-legume species with good efficacy performance. Application of a biofilmed inoculant containing a fungal rhizobia consortium greatly improved N_2 fixation in soybean relative to the conventional inoculant *Rhizobium*. Wheat seedlings inoculated with biofilm-producing bacteria showed an improved yield in moderately saline soils. Biofilms also tend to help microorganisms survive after inoculation under challenging circumstances; this is a crucial feature of the success of PGPM inoculation under agricultural circumstances. Biofilm inoculums have been shown to cause their rhizobia to live at high salinity (400 mM NaCl) by 105-fold, equivalent to *Rhizobium* monocultures. Interestingly, beneficial endophytes have been shown to develop higher acidity and growth-promoting hormones than their mono-or mixed cultures with no biofilm development [36].

Probiotic Metabolites

Prebiotics, as contrasted to antibiotics, are dietary supplements (tablets and capsules) and foods (yoghurt) that contain friendly bacteria [3]. It has become abundantly evident that the configuration of the intestinal microbiota and so, most of its skeletal muscle features can be influenced by comparatively modest dietary adjustments, such as the addition of prebiotics to enhance host health. A nutritional prebiotic is described as "a selectively fermented ingredient that causes particular changes in the behaviour of the gut tract, resulting in beneficial effects for the host" [37]. Bifidobacteria and lactobacilli are two beneficial bacterial classes that function as probiotic bacteria, which help in prebiotic metabolism [38, 39].

When provided in sufficient quantities, probiotic bacteria have a favourable impact on the host's health, while prebiotics are a class of dietary supplements that have a massive influence on the host by fostering colon growth and development and host health [40]. Synbiotic is a synergistic combination of probiotics and prebiotics [41, 42]. Prebiotics are dietary fibres that have been shown to have gastrointestinal benefits [43, 44].

Biosurfactants

Surfactors are amphiphilic microbial molecules, which are split into liquid/liquid, liquid/gas, and liquid/solid interfaces with hydrophilic and hydrophobic fractions. These characteristics enable certain biomolecules, which are important attributes in different industry sectors, to play an essential role in emulsion, foaming, and washing powder. Most of the new surfactants are extracted from petroleum chemically. But synthetic tension active agents are normally toxic and microorganisms can not be degraded. Microorganism surfactants known as "biosurfactants" are safer for the environment [45].

Microorganisms (bacteria and certain filamentous fungi) are capable of generating biosurfactants of various molecular structures and surface behaviours. Bacteria in the genera *Pseudomonas* and *Bacillus* are significant sources of biosurfactants. However, the majority of biosurfactants of bacterial origin are ineffective for use in the food industry due to their potential pathogenic nature. Various metabolic pathways are involved in the synthesis of precursors for the development of biosurfactants and rely on the existence of the major carbon sources used in the culture medium. Biosurfactants are formed by microorganisms either by excretion or by adhesion to cells. Microorganisms generate biosurfactants when they are grown on substrates that are insoluble in water. The key physiological function of such biosurfactants is to enable microorganisms to expand by reducing the surface tension between phases, making the substrate more usable for absorption and metabolism [45].

Antimicrobial Natural Products

Microorganisms such as bacteria and fungi create various compounds, which can work against other microorganisms (*e.g.* antimicroscopic SM's). Antimicrobial metabolites are used to prevent other species from vying for the same ecological niche through the aggressive development of biosynthetic organisms. Bacteria and fungi SM's are widely considered in medicinal usage to be antibiotics, such as penicillins, cephalosporins, tetracyclines, aminoglycosides, and chloramphenicol [24]. Antimicrobial drugs include all agents that function against different microorganisms-bacteria (antibacterial), viruses (antiviral) and fungi (anti-microbials).

Bacteriocins, proteinaceous antibacterial compounds commonly produced by many species of bacteria of lactic acid, as well as by both Gram ve- and Gram ve+ bacteria are a primary source of secondary antibiotic metabolites like fungus and actinomycetes [46, 47]. However, lactic acid bacteria produce lactic bacteria, which have a bacterial impact on a target location and are used as food preservatives [48]. The most common synthesized bacteriocin products are nisine, diplococcin, acidophiline, pediocin, bulgarian, helveticin, lactacin, and plantaricin [49].

Actinomycetes have been used in many different therapeutic fields and form the basis of the greatest number of potential antibiotic candidates and lead molecules. The members of order actinomycetals alone produced 45% of the currently recognized bioactive microbial metabolites; more than 10,000 compounds were already isolated from separate actinomycetes, 34% were streptomycetes, and 11% were non-streptomycetes [49, 50]. Actinomycetes produce essential classes of antibiotics, including β-lactams, aminoglycosides, lipopeptides, glycopeptides, asamycins, anthracyclines, nucleosides, peptides, polyethrines, tetracyclines, and macrolides.

CERTAIN IMPORTANT SOIL MICROBES AND THEIR INDUSTRIAL APPLICATIONS

The Global Footprint Network reported in 2018 that natural resources are consumed 1.7 times faster than Earth's recovery potential. In recent years, climate change, resource depletion, global warming, water shortages, and growing natural disasters have resulted in the degradation of water, soil, and air caused by anthropogenic intervention. Many researchers and projects are involved in the research and development of new methods, feedstocks, and processes for a sustainable economy less dependent on petroleum as an energy source and chemical processing.

The cost of operations (with feed acquisition) as well as the environmental effects of the inappropriate handling of residues can be minimized by using corn steep liqueur, energy crops, grass silage, bovine slurry, storage residues, and food production waste. This involves green materials that can be processed into precious biofuels (for example, biobutanol and bioethanol) and biochemicals (lactic acid, butyric acid, and succinic acid), which balance environmental sustainability and economic growth in a process called biorefining. "Sustainable biomass processing into a variety of commercializable foods and feed ingredients, bio-based products (chemical products, fabrics, *etc.*), and bioenergy (biofuels, electricity, and heat)" is the concept of bio-refining. Biomass materials are grouped into building blocks in a biorefinery (for example, cellulose, hemicellulose, lignin, and proteins) and the principle of biorefinery involves the whole management of biomass, taking into account upstream, midstream, and downstream production. The biggest challenge is to make biofuels and biochemicals a viable commodity with a cost-efficient and sustainable path. Though green feedstuffs are being used, the sustainability of biorefineries should also take note of the environmental and economic impacts of other factors, such as land use for biomass processing, soil and air quality during processing. Microbial cells are central to biomass conversion considering biochemical processes, acting as biocatalysts that convert liquid, solid, or gaseous substrates through a sequence of metabolic reactions in alcohols, acids, enzymes, and other materials.

Clostridium spp.

Clostridium spp. are typically in the form of rods with Gram-positive colouration. Single diameters of cells can vary between 0.3 and 2.0 μm and lengths of 1.5 to 20 μm are typically arranged with rounded or pointed ends in pairs or short chains. In the presence or absence of flagella, a single cell may be movable or immobile. They are not sulphate-reduction and catalase negative in general. Many animals have the ability to sporulate, which can be guided to an anaerobic oxygen presence. However, the oxygen tolerance ability of *Clostridium* species varies. The majority of animals are chemotrophic; some of them are chemotrophic or chemical. Saccharolytic, proteolytic, neither, nor either. They are metabolically very distinct, having an ideal temperature of 10°C to 65°C. They are scattered across the world. In particular, *Clostridium* species have been involved in many important fermentative processes, such as the production of butanol, acetone, ethanol, acetic acid, lactic acid, succinic acid, 1,3-propanediol, and more (Table 2).

Table 2. Main products of *Clostridium* species.

Species	Main Products
C. carboxidivorans	Ethanol, butyric acid, and butanol
C. beijerinckii	Butanol and isopropanol
C. saccharobutylicum	Butanol and ABE 5
C. saccharoperbutylacetonicum	Butanol, ABE 5, and acetone
C. acetobutylicum	Butanol, ethanol, and acetone
C. butyricum	1,3 - propanediol
C. pasteurianum	1,3 – propanediol and butanol

ABE: sum of acetone, butanol and ethanol concentrations.

Clostridium carboxidivorans

In the presence of carbon (CO_2 and CO), and also with various organic sources such as fructose, galactose, glucose, and mannose, this species has an autotrophic development. It contains finishing metabolic products like acetic acid, ethanol, butyrate, and butanol. In the genomic analysis of *C. carboxidivorans* to understand biofuel production pathways, genetic determinants were observed for CO use and production of acetate, ethanol, and butanol. One example is the presence of a gene encoding a dehydrogenase characteristic of the ABE (acetone-butanol-ethanol) fermentation, previously described for *C. acetobutylicum* and *C. beijerinckii*, responsible for converting acetyl-CoA to acetaldehyde and ethanol, and acetyl-CoA to butyryl-CoA and butanol *via* butyraldehyde. However, genes present in other *Clostridium* strains that encode proteins for acetone production were not found in *C. carboxidivorans*.

Clostridium beijerinckii

C. beijerinckii is a major bacteria containing isopropanol in many strains known to occur. It is also known as a butanol source (ATCC 25752, CIP 104308, DSM 791, JCM 1390, LMG 5716, and NCTC 13035). The species uses lignocellulosic hydrolysate products to obtain a biorefinery interest and produces butanol, and acetone, or isopropanol. Butanol is produced in large quantities from sugarcane sauté or cob hydrolysate [51]. This species produces alcohol, but only a few strains produce isopropanol-dehydrogenase, which converts acetone into isopropanol by fermentation with isopropanol-butanol-ethanol (IBE). Zn^{2+}, Ca^{2+} and Fe^{2+} have been reported to increase the production of butanol and isopropanol [51]. They also increase the basic hydrogenase function, which promotes the development of H_2.

Clostridium saccharobutylicum

Acetone, butanol, fructose, CO_2, and H_2 are produced as fermentation products by Saccharobutylicum, as well as acetic and butyric acids. It is primarily used in the manufacture of ABE (acetone-butanol-ethanol) fermented butanol, a maize stove, through lignocellulosic fermentations.

Clostridium saccharoperbutylacetonicum

As final fermentation products, *C. saccharoperbutylacetonicum* contains acetone, butanol, ethanol, CO_2, H_2, acetic acid, and butyric acid. It is considered an excellent model species to manufacture 73% - 85% butanol and, with low frequency, for the industrial use of acetone-butanol-ethanol (ABE) fermentation. *C. saccharoperbutylacetonicum* cell extractive fermentation contributed to elevated overall levels of butanol (64.6 g / L). The use of genetic modification methods helps explain genetic knowledge better. The strain *C. saccharoperbutylacetonicum* N1-4C, therefore, showed improved stability of the plasmid and small increases in the output of ABE, possibly due to an improved assimilation acid performance during the acidic stage.

The muzzling of pta and buk genes, responsible for the expression of phosphotransacetylase and butyrate kinase, did not remove acetyl or butyrate production, respectively.

Clostridium pasteurianum

It is capable of producing chemicals, such as 1,3-propanediol and n-butanol, from renewable raw materials, such as glycerol derived from biodiesel and glucose from the hydrolysis of biomass, respectively. Ethanol, acetic acid, butyric acid, lactic acid, H_2, and CO_2 are also among the products of its fermentation. Although necessary to ensure intracellular redox balance, the organic acid production pathway competes with that of the products of interest. There is also selectivity in the proportion of the products generated; for example, there is 1,3-propanediol and n-butanol, which is strongly influenced by factors such as nutrient composition and growing conditions. An increase in the production of butanol by *C. pasteurianum* DSM 525 was observed when lactic acid was used with glycerol.

Bacillus subtilis

There are various uses for bacteria that are probiotic. This is attributed to several potential health consequences of probiotics. Probiotics are available in powder, capsules, improved yogurts, yogurt-like products, and milk. Cholesterol reduction, immune system excitation, and the prohibition of cancer violence are the health consequences associated with them.

B. subtilis is an ideal multi-functional probiotic that not only hosts bioreactors but is also highly likely to prevent pathogenic bacteria from growing and to improve the nutrient absorption [52]. It is also used for vitamins, inositol, acetoin, hyaluronan, and other chemicals as an industrial cell factory. Its clear legacy history and well-developed gene manipulation tools have allowed its cell metabolism to rebuild, making it appealing to the public knockout collection as a metabolic engineering host [53].

Pseudomonas spp.

The use of *Pseudomonas* in the development of secondary metabolites from different pathways, which are equally useful in industry and medicine, will be discussed in this chapter. Pigments of *Pseudomonas* and their utility in bioelectricity, medicine, *etc.* will be considered.

Pseudomonas aeruginosa

P. aeruginosa provides a broad range of compounds essential for the treatment of bacteriostatically acting multi-drug-resistive (MDR) bacteria [54]. These molecules seem to be the outcome of secondary metabolism and are called SMs, such as the pathways of polyketide and shikimic chorismic acid [55]. Experiments have shown that substances produced naturally by microbes with antibacterial activity may be used in the diagnosis of diseases in humans, animals, and plants [56, 57]. Compounds with antimicrobial properties, including a group of peptides called pyocins and other heterocyclic compounds such as quinolines, phenylpyrroles, and phenazines, are developed by *Pseudomonas aeruginosa* strains. In aiding the colonisation of *P. aeruginosa,* these heterocyclic compounds remove microorganisms through DNA damage and cell depolarization [58].

P. aeruginosa has beneficial uses in various industrial and commercial sectors around the globe. These include waste degradation, oil refineries, textile products, agriculture, pulp and paper, mining, and explosive industries. Crude oil released into sea or land habitats is very hazardous and can harm the lives of those in the air, water, and soil systems, as well as pose the possibility of fire risks [59]. Onshore crude oil spills impact living ecosystem types, reducing farm productivity and polluting groundwater or potable water supplies and live biota, amongst other things, in flowing water bodies [60]. The removal of these harmful effects from crude oil discharges implies, wherever possible, complete prevention and soil modification through the so-called bioremediation process [61].

Aspergillus spp.

A. niger is an important fungal species used in the processing of food additives, extracellular enzymes, and organic acids [62] and in biotechnology [63]. Numerous strains of *A. Niger* are deemed by the United States Food and Drug Administration (USFDA) to be widely accepted as safe (GRAS) [64]. Consequently, A. Niger has been used in the mass processing of homologous or heterologous proteins for decades as a host mechanism [65]. The latest research, however, finds that some *A. niger* strains can generate damaging secondary metabolites, including ochratoxin A and fumonisin. *A. oryzae* and *A. Niger* each produce a considerable quantity of extracellular enzymes, and amylases are of significant industrial importance [66].

A. fumigatus can produce pyripyropene, which is another cholesterol lowering agent that inhibits acyl-CoA: cholesterol acyltransferase. A well-known antibiotic, penicillin, is produced by several *Aspergillus* species, including *A. nidulans* [67]. *Aspergillus* species also produce other products that have found clinical utility as anticancer or antifungal agents [68].

CONCLUDING REMARKS

The research on the substitution of chemical compounds with green biomaterials has grown in response to concerns about the supply of organic, nutritious diets for humans. Different methods of making and treating sustainable monomers and polymers, taking into account their advantages and drawbacks, are currently being researched. For food, medication, and medicinal uses, there is a need to create a green alternative. Due to the simplicity of the mechanism and transformation into essential PMs and SMs, microbial synthesis, for instance, biosynthesis, has attracted great interest. There are many benefits to the approach of green chemistry for the synthesis of natural materials by microorganisms, such as the simplicity with which a method can be extended, along with economic feasibility and health protection. The use of natural products in food, medicine, and agricultural sectors is extensively emphasised because of their chemical stability and bioactivity. Microbes have been used for many decades in the production of a wide range of biochemicals, from alcohol to antibiotics, as well as in food and feed manufacturing. However, the perspective on microbial synthesis is far broader as more resources are provided for producing the ultimate green biopolymers, biomaterials, biosurfactants, microbial antibiotics, antimicrobials, and food-coating biofilms.

CONSENT FOR PUBLICATION

Not applicable.

CONFLICT OF INTEREST

The author declares no conflict of interest, financial or otherwise.

ACKNOWLEDGEMENTS

Declared none.

REFERENCES

[1]   Bruce A, Alexander J, Julian L, *et al.* Eds. Molecular Biology of the Cell.. New York: Garland Science 2007; p. 1392.

[2]   Gupta C, Prakash D, Gupta S. Natural useful therapeutic products from microbes. J Microbiol Exp 2014; 1: 30-7.

[3]   Abdel-Aziz SM. Industrial Microbiology and Biotechnology. Garg N, Aeron A, Eds. Microbes in Process. USA: Nova Science Publishing Inc 2014; pp. 1-22.

[4]   Crueger W, Crueger A, Eds. Biotechnology: A Textbook of Industrial Microbiology. Stamford, Connecticut: Sinauer Associates 1990; p. 357.

[5]   Abdel-Aziz SM, Abo Elsoud MM, Anise AAH. Microbial Biosynthesis: A repertory of Vital Natural Products. In: Grumezescu AM, Holban AM, Eds. Food Biosynthesis: Handbook of Food Bioengineering. London, UK: Academic Press 2017; Vol. 1: pp. 25-54.

[6]   Bérdy J. Bioactive microbial metabolites: a personal view. J Antibiot (Tokyo) 2005; 58: 1-26.
        [PMID: 15813176]

[7]   Purves K, Macintyre L, Brennan D, *et al.* Using molecular networking for microbial secondary metabolite bioprospecting. Metabolites 2016; 6: 2-19.
        [PMID: 26761036]

[8]   Wohlgemuth R. Biocatalysis: key to sustainable industrial chemistry. Curr Opin Biotechnol 2010; 21: 713-24.
        [PMID: 21030244]

[9]   Ran N, Zhao L, Chen Z, *et al.* Recent applications of biocatalysis in developing green chemistry for chemical synthesis at the industrial scale. Green Chem 2008; 10: 361-72.

[10]   Adrio JL, Demain AL. Microbial enzymes: tools for biotechnological processes. Biomolecules 2014;

4: 117-39.
[PMID: 24970208]

[11] Chukwuma C. Control mechanisms of protein-protein interactions in biosynthetic pathways of bacteria and fungi. Appl Sci Res 2016; 14: 258-67.

[12] Barker HA. Amino acid degradation by anaerobic bacteria. Annu Rev Biochem 1981; 50: 23-40.
[PMID: 6791576]

[13] Davila A-M, Blachier F, Gotteland M, *et al.* Re-print of "Intestinal luminal nitrogen metabolism: role of the gut microbiota and consequences for the host". Pharmacol Res 2013; 69: 114-26.
[PMID: 23318949]

[14] Brautaset T, Ellingsen TE, Eggeling L. Lysine Industrial Uses and Production.Moo-Young M, Ed. Comprehensive Biotechnology. 2nd ed. Amsterdam, The Netherlands: Elsevier 2011; pp. 541-54.

[15] Maiorella BL, Castillo FJ. Ethanol, biomass, and enzyme production for whey waste abatement. Process Biochem 1984; 19: 157-61.

[16] Taherzadeh MJ, Lennartsson PR, Teichert O, *et al.* Bioethanol Production Processes.Babu V, Thapliyal A, Patel GK, Eds. Biofuels Production. New Jersey: John Wiley & Sons, Inc 2013; pp. 211-53.

[17] Dürre P. New insights and novel developments in clostridial acetone/butanol/isopropanol fermentation. Appl Microbiol Biotechnol 1998; 49: 639-48.

[18] Zverlov VV, Berezina O, Velikodvorskaya GA, *et al.* Bacterial acetone and butanol production by industrial fermentation in the Soviet Union: use of hydrolyzed agricultural waste for biorefinery. Appl Microbiol Biotechnol 2006; 71: 587-97.
[PMID: 16685494]

[19] Chiao J-S, Sun Z-H. History of the acetone-butanol-ethanol fermentation industry in China: development of continuous production technology. J Mol Microbiol Biotechnol 2007; 13: 12-4.
[PMID: 17693708]

[20] Angumeenal AR, Venkappayya D. An overview of citric acid production. Lebensm Wiss Technol 2013; 50: 367-70.

[21] Carlos RS, Luciana PS, Vandenberghe RC, *et al.* Citric acid production. Food Technol Biotechnol 2006; 44: 141-9.

[22] Ige O, Lasisi E, Aribo S. Natural products: a minefield of biomaterials. ISRN Mater Sci 2012; 2012: 983062.
[http://dx.doi.org/10.5402/2012/983062]

[23] Li B, Sher D, Kelly L, *et al.* Catalytic promiscuity in the biosynthesis of cyclic peptide secondary metabolites in planktonic marine cyanobacteria. Proc Natl Acad Sci USA 2010; 107: 10430-5.
[PMID: 20479271]

[24] Weber T, Kim HU. The secondary metabolite bioinformatics portal: Computational tools to facilitate synthetic biology of secondary metabolite production. Synth Syst Biotechnol 2016; 1: 69-79.
[PMID: 29062930]

[25] Tianero MD, Pierce E, Raghuraman S, *et al.* Metabolic model for diversity-generating biosynthesis. Proc Natl Acad Sci USA 2016; 113: 1772-7.
[PMID: 26831074]

[26] Hernández N, Williams RC, Cochran EW. The battle for the "green" polymer. Different approaches for biopolymer synthesis: bioadvantaged *vs.* bioreplacement. Org Biomol Chem 2014; 12: 2834-49.
[PMID: 24687118]

[27] Wani SJ, Shaikh S, Sayyed RZ. Microbial biopolymers in biomedical field. MOJ Cell Sci Rep 2016; 3: 65-7.

[28]   Schmid J, Sieber V. Enzymatic transformations involved in the biosynthesis of microbial exo-polysaccharides based on the assembly of repeat units. ChemBioChem 2015; 16(8): 1141-7.
[PMID: 25873567]

[29]   Adsul M, Tuli DK, Annamalai PK, *et al.* Polymers from biomass: characterization, modification, degradation, and applications. Int J Polym Sci 2016; 2016: 1857297.
[http://dx.doi.org/10.1155/2016/1857297]

[30]   Eldin MSM, Soliman EA, Hashem AI, *et al.* Biopolymer Modifications for Biomedical Applications.Theophile T, Ed. Infrared Spectroscopy—Life and Biomedical Sciences InTech open. 2012; pp. 355-68.

[31]   Shit SC, Shah PM. Edible polymers: challenges and opportunities. J Polym 2014; 2014: 427259.
[http://dx.doi.org/10.1155/2014/427259]

[32]   Okaiyeto K, Nwodo UU, Okoli SA, *et al.* Implications for public health demands alternatives to inorganic and synthetic flocculants: bioflocculants as important candidates. MicrobiologyOpen 2016; 5: 177-211.
[PMID: 26914994]

[33]   Ugbenyen AM, Okoh AI. Characteristics of a bioflocculant produced by a consortium of *Cobetia* and *Bacillus* species and its application in the treatment of waste waters. Water SA 2014; 40: 139-44.

[34]   Wang M, Kuo-Dahab WC, Dolan S, *et al.* Kinetics of nutrient removal and expression of extracellular polymeric substances of the microalgae, *Chlorella* sp. and *Micractinium* sp., in wastewater treatment. Bioresour Technol 2014; 154: 131-7.
[PMID: 24384320]

[35]   Qureshi N, Annous BA, Ezeji TC, *et al.* Biofilm reactors for industrial bioconversion processes: employing potential of enhanced reaction rates. Microb Cell Fact 2005; 4: 24.
[http://dx.doi.org/10.1186/1475-2859-4-24] [PMID: 16122390]

[36]   Bandara WMMS, Seneviratne G, Kulasooriya SA. Interactions among endophytic bacteria and fungi: effects and potentials. J Biosci 2006; 31: 645-50.
[PMID: 17301503]

[37]   Gibson GR, Scott KP, Rastall RA, *et al.* Dietary prebiotics: current status and new definition. Food Sci Technol Bull: Function Foods 2010; 7: 1-19.

[38]   Gibson GR, Rastall RA, Roberfroid MB. Prebiotics. Gibson GR, Roberfroid MB, Eds. Colonic Microbiota, Nutrition and Health. Doordrecht, The Netherlands: Kluwer Academic Press 1999; pp. 101-24.

[39]   Bouhnik Y, Raskine L, Simoneau G, *et al.* The capacity of nondigestible carbohydrates to stimulate fecal bifidobacteria in healthy humans: a double-blind, randomized, placebo-controlled, parallel-group, dose-response relation study. Am J Clin Nutr 2004; 80: 1658-64.
[PMID: 15585783]

[40]   Roberfroid MB. Prebiotics and probiotics: are they functional foods? Am J Clin Nutr 2000; 71: 1682S-7S.
[PMID: 10837317]

[41]   Gibson GR, Roberfroid MB. Dietary modulation of the human colonic microbiota: introducing the concept of prebiotics. J Nutr 1995; 125: 1401-12.
[PMID: 7782892]

[42]   Vyas U, Ranganathan N. Probiotics, prebiotics, and synbiotics: gut and beyond. Gastroenterol Res Pract 2012; 2012: 872716.
[http://dx.doi.org/10.1155/2012/872716] [PMID: 23049548]

[43]   de Vrese M, Marteau PR. Probiotics and prebiotics: effects on diarrhea. J Nutr 2007; 137: 803S-11S.
[PMID: 17311979]

[44]   Raman M, Ambalam P, Kondepudi KK, *et al.* Potential of probiotics, prebiotics and synbiotics for management of colorectal cancer. Gut Microbes 2013; 4: 181-92.
[PMID: 23511582]

[45]   Santos DK, Rufino RD, Luna JM, *et al.* Biosurfactants: multifunctional biomolecules of the 21st century. Int J Mol Sci 2016; 17: 401-32.
[PMID: 26999123]

[46]   Deshmukh PV, Thorat PR. Detection of antimicrobial efficacy of novel bacteriocin produced from *Lactobacillus similis* RL7. Int J Adv Res (Indore) 2014; 2: 987-95.

[47]   Savadogo A, Ouattara ATC, Bassole HNI, *et al.* Bacteriocins and lactic acid bacteria: a mini review. Afr J Biotechnol 2009; 5: 679-84.

[48]   Hafidh RR, Abdulamir AS, Vern LS, *et al.* Inhibition of growth of highly resistant bacterial and fungal pathogens by a natural product. Open Microbiol J 2011; 5: 96-106.
[PMID: 21915230]

[49]   Hayek S, Rabin G, Salam A. Antimicrobial Natural Products. Méndez-Vilas A, Ed. Microbial Pathogens and Strategies for Combating Them: Science, Technology, and Education. Spain: Formatex Research Center 2013; pp. 910-21.

[50]   Adegboye M, Babalola O. Actinomycetes: a yet Inexhaustive Source of Bioactive Secondary Metabolites. Méndez Vilas A, Ed. Microbial Pathogens and Strategies for Combating Them: Science, Technology, and Education. Spain: Formatex Research Center 2013; pp. 786-95.

[51]   Zhang J, Jia B. Enhanced butanol production using *Clostridium beijerinckii* SE-2 from the waste of corn processing. Biomass Bioenergy 2018; 115: 260-6.

[52]   Olmos J, Acosta M, Mendoza G, *et al. Bacillus subtilis*, an ideal probiotic bacterium to shrimp and fish aquaculture that increase feed digestibility, prevent microbial diseases, and avoid water pollution. Arch Microbiol 2020; 202: 427-35.
[PMID: 31773195]

[53]   Gu Y, Xu X, Wu Y, *et al.* Advances and prospects of *Bacillus subtilis* cellular factories: From rational design to industrial applications. Metab Eng 2018; 50: 109-21.
[PMID: 29775652]

[54]   Butler MS, Blaskovich MA, Cooper MA. Antibiotics in the clinical pipeline in 2013. J Antibiot (Tokyo) 2013; 66: 571-91.
[PMID: 24002361]

[55]   Bérdy J. Bioactive microbial metabolites. J Antibiot (Tokyo) 2005; 58: 1-26.
[PMID: 15813176]

[56]   Depoorter E, Bull MJ, Peeters C, *et al.* Burkholderia: an update on taxonomy and biotechnological potential as antibiotic producers. Appl Microbiol Biotechnol 2016; 100: 5215-29.
[PMID: 27115756]

[57]   Thi QV, Tran VH, Mai HD, *et al.* Secondary Metabolites from an Actinomycete from Vietnam's East Sea. Nat Prod Commun 2016; 11: 401-4.
[PMID: 27169191]

[58]   Onbasli D, Aslim B. Determination of antibiotic activity and production of some metabolites by *Pseudomonas aeruginosa* B1 and B2 in sugar beet molasses. Afr J Biotechnol 2008; 7: 4814-9.

[59]   Abioye O. Biological Remediation of Hydrocarbon and Heavy Metals Contaminated Soil. Pascucci S, Ed. Soil Contamination. Intech Open 2011; pp. 127-42.

[60]   Perelo LW. Review: *In situ* and bioremediation of organic pollutants in aquatic sediments. J Hazard Mater 2010; 177: 81-9.
[PMID: 20138425]

[61]   Chhatre S, Purohit H, Shanker R, *et al.* Bacterial consortia for crude oil spill remediation. Sci Technol 1996; 34: 187-93.

[62]   Pandey A, Selvakumar P, Soccol CR, *et al.* Solid state fermentation for the production of industrial enzymes. Curr Sci 1999; 77: 149-62.

[63]   Schuster E, Dunn-Coleman N, Frisvad JC, *et al.* On the safety of *Aspergillus niger*-a review. Appl Microbiol Biotechnol 2002; 59: 426-35.
[PMID: 12172605]

[64]   Taylor MJ, Richardson T. Applications of microbial enzymes in food systems and in biotechnology. Adv Appl Microbiol 1979; 25: 7-35.
[PMID: 397740]

[65]   Archer D, Turner G. Genomics of Protein Secretion and Hyphal Growth in *Aspergillus*. Brown AJ, Ed. Fungal Genomics The Mycota . Berlin, Heidelberg: Springer 2006; Vol. 13: pp. 75-96.

[66]   Hernandez MS, Rodriguez MR, Guerra NP, *et al.* Amylase production by *Aspergillus niger* in submerged cultivation on two wastes from food industries. J Food Eng 2006; 73: 93-100.

[67]   Dulaney EL. Some aspects of penicillin production by *Aspergillus nidulans*. Mycologia 1947; 39: 570-81.
[PMID: 20264542]

[68]   Bladt TT, Frisvad JC, Knudsen PB, *et al.* Anticancer and antifungal compounds from *Aspergillus, Penicillium* and other filamentous fungi. Molecules 2013; 18: 11338-76.
[PMID: 24064454]

CHAPTER 3

Industrial Aspects of Microbes

B.J. Yogesh[1,*] and **S. Bharathi**[1]

¹ Department of Microbiology, The Oxford College of Science, Bangalore, India

Abstract: This chapter deals with the significance of soil microbes from an industrial perspective. Soil microbes are the most diverse populations to exist on earth, and they are known to have played a prominent role in the development of soil chemistry, soil texture, and soil suitability to sustain plant life. The chapter deals with the significance of cultural techniques for the isolation of desired microbial strains from the soil. The importance of screening techniques for isolates is emphasized, wherein the potential strains are tested for their physiological characteristics that are industrially beneficial. A few criteria are mentioned for judging the soil isolate's capability to become an industrial strain. The difference between natural isolates and potential industrial strains is discussed. Useful strains are categorized based on their ability to produce primary and secondary metabolites with commercial applications in terms of economic, agricultural, and environmental significance. Industrially important microbes are listed with emphasis on the types of metabolites they produce and their applications. Knowledge of metabolic pathways involved in metabolite production and their regulation in terms of various feedback control systems are discussed. Strain improvement and its role in improving industrial aspects of microbes are highlighted. *Bacillus* sp. are given their due importance as the most diverse and dynamic forms of bacteria, contributing immensely to our knowledge and being the most beneficial forms of soil microbes. A few metabolites are discussed in detail, with emphasis given to enzymes, microbial polymers, amino acids, solvents, organic acids, and antibiotics. Microbial bioleaching mostly employs bacteria that could help in the recovery of metals from low-grade ores, and industries based on biomining have shown a renewed interest in this economically viable process.

Keywords: Antibiotics, Isolates, Metabolites, Organic acids, Strain selection, Solvents.

INTRODUCTION

Soil is a reserve of diverse microbial populations, ranging from single-celled bacteria to complex fungi and algal forms that make soil what it is- a growth hub for plants and a recycling sink for dead organic matter. The present scenario of

---

* **Corresponding author B.J. Yogesh:** Department of Microbiology, The Oxford College of Science, Bangalore, India; E-mail: jyogesh2009@gmail.com

Ashutosh Gupta, Shampi Jain & Neeraj Verma (Eds.)

excessive environmental and soil pollution has made the scientific community acknowledge the significance of soil microbes and their positive impact on soil dynamics. Soil microbes are getting renewed interest and are increasingly viewed as an important partner in sustainable agricultural practices. Now, organic farming is preferred as a safer means of avoiding the use of chemical fertilizers, and microbial biocontrol agents are being recommended to replace chemical insecticides and pesticides.

Efficient nitrogen-fixing microbes are marketed and have found new stakers among farming communities. Plant-microbe interactions have given a new perspective on the role of microbes in plant growth with their ability to produce diverse metabolites like plant growth-promoting factors, antibiotics, siderophores, enzymes, antifungal molecules, anti-parasitic agents, antiviral drugs, other bio-pharmaceuticals of industrial relevance, *etc*. These studies have given a new dimension to soil microbes as a source of potential metabolites of industrial significance.

Soil microorganisms are in a constant struggle for their survival and thus have adapted to this competitive environment by being sturdy, tightly regulated, and fine-tuned in metabolism, adapted to starvation and stress, having efficient metabolite transport systems, diverse enzymes to degrade a wide range of substrates, and producing a series of secondary metabolites to outcompete other soil microflora. Slow-growing actinomycetes and fungi have thus unleashed a series of antibacterial compounds as part of their survival strategy. Spore-forming bacteria like *Bacillus* and *Clostridium* would wait longer for favourable conditions to return. Microbes cling to the rhizosphere region for better survival, and some forms have developed a symbiotic relationship with plants. Most microbes have adapted to the role of decomposers and have developed an array of enzymes in this regard. A few of them have taken the path of pathogens to attack plants for their sustenance. Hence, the soil is a complex zone of microbial interactions where we could come across diverse life forms, each adapted to existence with its own self-survival technique, which is unique and rewarding to the concerned microbe.

Microbial type and load vary with soil and are influenced by the existing soil profile, soil texture and its content, and soil interaction with the immediate atmosphere. For example, soil texture determines moisture retention capacity and, when compared to annual precipitation rate, may determine whether the soil is too dry (suitable for actinomycetes), moistened, and well aerated (suitable for diverse bacterial and fungal loads), or waterlogged (allows algae to dominate). Thus, physical parameters influence microbial diversity in soil.

Similarly, soil exposed to dead and decaying matter may be super rich in decomposers; not so well aerated soil may carry diverse anaerobes; soil exposed to an oil spill and petroleum products (near refineries and petrol stations) would have hydrocarbon-degrading microbes. All the above assessments are based on long-time exposure of microbes to particular physical conditions that would lead to microbial adaptation and evolution.

Different methods of assessing microbial types and biomass from soil have been developed, ranging from cultural methods to molecular techniques (genetic and immunological). Soil can be analyzed for total microbial load by plate count analysis and the membrane filtration method. Microbe types and their numbers may vary, so dilution is essential. The bacterial load is far greater than previously thought, owing to faster growth rates and greater adaptability to diverse nutrients in the soil, followed by actinomycetes and mould populations. Nutrient agar, glycerol yeast extract agar, and glucose peptone acid agar are preferred media for bacteria, actinomycetes, and fungi, respectively. A direct microscopic count can give a quantitative measure of total microbial populations. Real-time analysis of soil microbes is based on molecular techniques like real-time-polymerase chain reaction (RT-PCR), random primers based RAPD, denatured gradient gel electrophoresis, various immunological tests, an assay of biomass components (DNA, protein, ergosterol, and glucosamine estimation), and an assay of metabolic activities. Some of the techniques mentioned above are indirect assays that may be useful in assessing total microbial load but will not provide the precise type of microbe. Drawbacks are in the form of dead and decaying matter, which may add up to the values.

PROCEDURE FOR ISOLATION OF MICROBES FOR COMMERCIAL PURPOSE

Isolation of Desired Strain

Basic microbiological techniques involving proper soil sample collection, transport, and appropriate storage of samples, the extent of serial dilution, types of media prepared, incubation conditions, and identification of isolates are considered classical and time-tested techniques. However, a prerequisite to all this is what you are searching for. The entire process depends on your needs. Without a preplanned of your exact needs, you would end up handling a huge pile of microbial colonies, which would be so exhaustive that it would be impossible to attain your goal. For example, if you are looking for a microbe that produces cellulase enzyme, methane, or iron-reducer, you will need to narrow your search. For isolating cellulase producers, it would be appropriate to choose an organic-rich garden or composting soil, and a methanogen can be isolated from an anaerobic digester or water-logged paddy fields, while an iron-reducing bacteria can be searched for in the subsurface or subterranean soil. Thus, the target of isolating a potential strain is half solved by choosing the right sampling point, while the remaining problem can be solved by devising selection techniques that preferably isolate desired strains.

Selection of Useful Strains

There is an unending search for prospective soil microbes that have significant industrial applications. Most of the industrially important microbes are directly sourced from soil and later made suitable for industry *via* strain improvement. Isolation of a suitable microbial strain requires screening, but it is rather impossible to mimic the soil environment in the laboratory. Hence, microbiologists end up isolating only a handful of microbes, and that too, rather than the same species again and again. In this regard, researchers have adopted innovative screening techniques that benefit the growth of less studied forms of microbes. Screening of soil microbes emphasizes providing enrichment media, which would allow the growth of a particular type of microbe while inhibiting unwanted and undesired forms. The enrichment media is essential, especially in limiting the growth of dominant and contaminating microbes, thus paving the way for lesser-known types present in the soil. Certain inhibitors help in controlling the growth of bacterial types while allowing the growth of slow growers like fungi and actinomycetes.

The selection of suitable strains depends on their growth abilities and adaptability to laboratory conditions. A few criteria are earmarked for consideration while selecting microbial cultures, such as:

- The strain should be stable enough to be repeatedly sub-cultured in lab conditions.
- It should be genetically stable and have a distinct morphology with consistency.
- It should grow faster and not be susceptible to contamination.
- It should prefer to grow in a simple nutritional environment.
- It has a wider range of preferences than physical environmental parameters.
- It produces a metabolite of industrial significance.
- End product accumulation is metabolically efficient.

Screening of Soil Isolates

The screening of the selected strain is carried out to confirm that the above criteria have truly been attained. Screening also tests the microbes in different stress conditions, which is similar to the soil environment for assessing their physiological behavior. Physical parameters like pH, temperature, oxygen tension, salts, and nutritional diversity in terms of carbon, nitrogen, and micronutrient influences are observed. Metabolic pathways and their regulations are investigated to understand their energy efficiency and adaptability to modifications. All the above analysis is essential to check the suitability of an isolate for industrial production. Further, screening of the selected strain has to be done to check adaptation to large scale fermentation in terms of utilizing cheaper nutritional sources, nonsusceptibility to infections, higher product yield, easier product purification, and economically feasible downstream processing. Appropriate care has to be taken to ensure that metabolically engineered organisms do not undergo a genetic reversal. In terms of ecological safety, genetically modified organisms should be contained and prevented from escaping into the environment. Pathogens from soil sometimes exhibit useful industrial traits, and in such cases, safety protocols have to be followed for handling, minimizing, or completely avoiding workers' exposure to strain.

Strain Improvement

A natural microbial candidate would have its limitations in terms of feedback control systems that cap the metabolic pathways from constitutive production of either primary or secondary metabolites. In this regard, it is essential to metabolically engineer the microbes by targeting their genes that are directly or indirectly responsible for end product formation. Strain improvement deals with the mechanism to achieve the goal of overproduction of products. Conventional mutation techniques are preferred as they can target the genetic mutation randomly and specifically too. Improvement of strains not only helps in achieving overproduction of end products but also on an array of other features of the microbial culture that would help in overall performance in terms of upstream and downstream processes. Industrially successful soil microbes are generally fast-growing, stable (physiologically, morphologically, and genetically), dominant, non-fastidious, scalable, amenable to mutation studies, quite adaptable to environmental changes, and well regulated.

INDUSTRIALLY IMPORTANT MICROBES

Bacteria dominate the soil by their sheer numbers and types, and are the most diverse of all life forms on earth, as soil holds an enormously uncharacterized bacterial population. Soil bacteria are known to produce a variety of significant primary and secondary metabolites, and as a result, researchers always look to the soil for solutions to their problems, whether they are agricultural, health-related, or industrial (Fig. **1**).

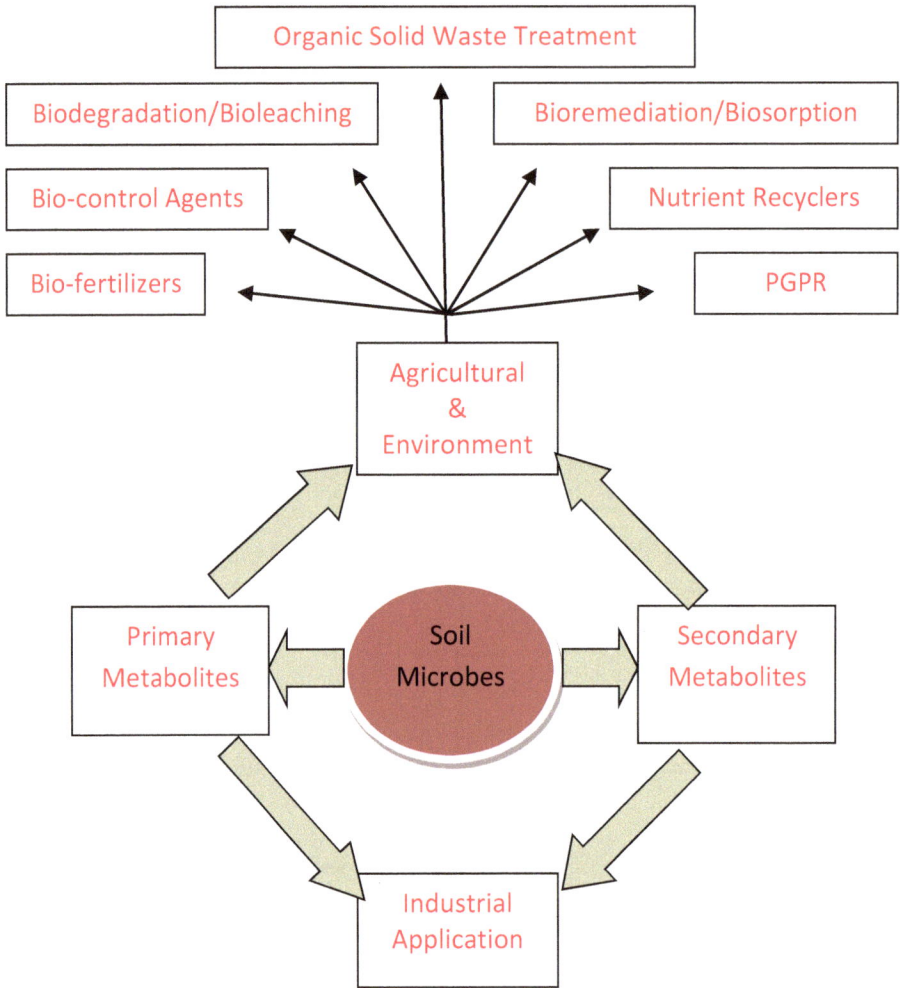

Fig. (1). Applications of soil microbes.

Aerobic bacteria dominate the soil surface and undergo the respiration mode of metabolism, using oxygen as a terminal electron acceptor. These aerobic bacteria tend to synthesize a wide range of primary metabolites required for growth, and sometimes these metabolites can leak out of the cell and thus create a microenvironment of diverse life forms. Microbial interactions lead to both positive and negative outcomes for the concerned microbe. This adaptation to competition has diversified bacteria to produce a host of secondary metabolites of significance. Such evolution has led to complex associations in terms of plant-microbe symbiosis, *e.g.*, *Rhizobium*.

Among bacteria, the most abundant form to dominate soil is *Bacillus* sp., which is a rich source of diverse primary and secondary metabolites of commercial significance (Table **1**). They are common inhabitants of the soil environment with the ability to decompose a wide range of organic matter. Though primarily identified as decomposers, they have occasionally been found to interact with plants and animals as well. Rod-shaped, often motile, chemoorganotrophic, and heterogeneous groups can be found in a variety of soil habitats and have a wide range of physiological responses to heat, pH, and salinity. They play an ecologically important role, chiefly attributed to their ability to withstand adverse environmental conditions through spore formation. A few species are pathogenic, but most of them are beneficial both agriculturally and industrially in terms of their metabolic activity and metabolites.

Table 1. Industrial significance of soil bacteria.

S. No.	Bacteria	Metabolite	Industrial Relevance
1.	*Bacillus*	Thuringiensin, zwittermicin A, bacillibactin, lipopeptide antibiotics, siderophores, amylases, and proteases	Commercial producers of enzymes. Antifungal agents, bacteriocin, insecticidal, biosurfactants, and decomposers
2.	*Pseudomonas*	Pseudobactin, pyochelin, pyoverdine, auxin, phenazine, pyrrolnitrin, HCN, oomycin, chitinases, and glucanases	Bio-control agents, phytohormones, antifungal agents, antinematodal, and antibacterial (cell wall degrading) enzymes
3.	*Rhizobium, Bradyrhizobium, Azorhizobium, Allorhizobium, Sinorhizobium,* and *Mesorhizobium*	Nod factors, rhizobitoxine, lytic enzymes, IAA, phosphate solubilizer, antibiotics, and siderophore	Commercial bioinoculant, Symbiont nitrogen fixer of commercially important crops, mineral solubilization, mobilization of nutrients, and bio-control agents
4.	*Azotobacter*	Auxins, IAA, gibberellic acid, biotin, nicotinic acid, pantothenic acid, and other B vitamins	Commercial bioinoculant, free living N_2 fixer, vitamins, and antifungal antibiotics
5.	*Azospirillum*	Antioxidant enzymes and phytohormones	Commercial bioinoculant and mitigation of abiotic stress (drought and salinity)
6.	Phosphate solubilizing bacteria	Organic acids, antibiotics, plant growth hormones, and siderophores	Commercial bioinoculant, soil aggregator, and resistance to abiotic stress

(Table 1) cont.....

S. No.	Bacteria	Metabolite	Industrial Relevance
7.	Aerobic bacteria	Enzymes involved in the degradation of pesticides, hydrocarbon, and PAHs	Soil bioremediation and herbicide producer
8.	Anaerobic bacteria	Enzymes involved in the degradation of PCBs	Soil bioremediation and herbicide producer
9.	Methylotrophs	Methane oxidizing enzymes, degradation of chlorinated aliphatics, and degradation of trichloroethylene	Soil bioremediation
10.	*Streptomyces*	Anthracyclines, avermectin, bafilomycin, arenimycin, bisanthraquinone, hygromycin, lincomycin, mitomycin, streptomycin, tetracycline, and sesquiterpene	Antitumor, antibacterial, anticancer, antiparasitic, immnosuppressive, diabetogenic, and cytotoxic agents

Many strains of *Bacillus* spp. are known to produce antimicrobial peptides and are yet to be tapped for their potential benefits. The bacteria have high survival chances in soil, equally attributed to their spore formation ability under adverse conditions and also to their being better competitors in terms of the secondary metabolites of antimicrobial nature.

Bacillus spp. are easier to handle in the lab, as they are fast-growing, preferably grown in less nutritious media, abundantly present in the soil, spore-forming, easy to manipulate in terms of strain improvements, high yield of end products, and scalability, and thus can be commercially exploited. *Bacillus* spp. have got a renewed interest in pharmaceutical industries for their yield of particular antibiotics, lipopeptide antibiotics (surfactants) with their surface active properties, and a series of enzymes in terms of amylases, neutral and alkaline proteases, with applications in environmental monitoring too.

Thousands of tonnes of atmospheric nitrogen can be effectively tapped for plant growth if nitrogen-fixing bacteria are employed as they do the process biologically, which is safer in comparison to chemical fertilizers that pollute the environment. *Rhizobium* and *Bradyrhizobium* fix nitrogen by forming nodules in plant roots. Legumes are abundant plant species with economic benefits to mankind. *Rhizobium* spp. can be isolated from such root nodules using yeast extract mannitol agar (YEMA) and found to be host specific, as different plant nodules yield different *Rhizobium* strains. Inoculant production can be done on a commercial scale using a cheap media source and then applied to the soil. Production is economically viable because it is a source of nutrition for growth, can be easily preserved and revived in times of need, and can be scaled up and cultivated economically. *Rhizobium* sp. has been found to produce several beneficial metabolites that are known to promote plant growth [1]. The fermentation process involves the employment of either solid state fermentation or a submerged type of fermentation and is finally formulated with a solid or liquid base. The commercially useful bioinoculants have been found to have additional properties that go beyond nitrogen fixation [2, 3]. Two important *Pseudomonas* spp., namely, *P. fluorescens* and *P. aeruginosa*, have been found relevant for beneficial exploitation and are known for their diverse metabolites of plant growth-promoting factors such as pseudobactin, pyochelin, pyoverdin, auxin, chitinases, and glucanases, to mention a few [4].

The global biofertilizer market is set to grow tremendously as eco-friendly soil management alternatives create a sustainable environment for viable agriculture. There is a need to cut dependence on chemical fertilizers as the damage to the environment is immense. Soil runoff has led to pollution of water bodies and eutrophication, negatively impacting aquatic life. In this regard, *Rhizobium, Azotobacter, Azospirillum, Glucobacter, Arthrobacter, Serratia, Xanthomonas, Enterobacter, Clostridium, Pseudomonas, Flavobacterium*, and *Bacillus* spp. have been found with defined values and importance (Table **2**).

Table 2. Commercially important metabolites of fungi.

S. No.	Fungi	Industrially Relevant Metabolites
1.	EM, AM, and VAM	Commercial bioinoculants
2.	*Aspergillus fumigatus*	Organic acids
3.	*Aspergillus niger*	Citric acid
4.	*Fusarium* sp. and *Penicillium* sp.	Several primary metabolites
5.	*Claviceps purpurea*	Ergot alkaloids
6.	*Gibberella fujikuroi*	Gibberellic acid
7.	*Aspergillus terreus*	Lovastatin (polyketides)

Fungi are an important micro-flora of soil and are next in importance to soil fertility and plant growth. They are immensely beneficial to soil in terms of nutrient recycling, decomposition of dead and decaying matter, and are host to a series of primary and secondary metabolites [5]. Primary metabolites of significance are enzymes of environmental and industrial significance. They degrade a series of plant polymers like cellulose and lignin. They are host to well-documented secondary metabolites with antimicrobial properties that have found commercial applications. Fungi, in particular, have a huge role in soil fertility in terms of plant associations that lead to nitrogen fixation, mobilization of minerals and solubilizing nutrients, and making them available to plants for growth [6]. Fungi are also host to bioremediation processes, composting, and can clean up various pollutants from the soil environment [7].

Large scale cultivation of fungi for the above products can be carried out either by solid state fermentation (SSF) or by submerged fermentation. Submerged fermentation is viable for large scale fermentation but comes with cost inputs, while solid state fermentation gives higher yield, less stringent control systems, and lower economic costs, but would most probably work out at pilot scale, where labour costs are higher. Physiologically, a fungus prefers SSF, as the moisture content is less and less of a competition in terms of contaminating bacteria. The proliferating fungi may achieve high density growth and a good yield of products. Monitoring fungal growth is an issue in SSF as the fungal biomass is closely knitted with the substrate, and hence indirect methods of monitoring microbial growth have been recommended.

Mycorrhizal fungi, including ectomycorrhizae (EM), endomycorrhizae (AM) and vesicular arbuscular mycorrhizae (VAM), are beneficial, lead to an increase in the uptake of nutrients, especially phosphorus, and exert a synergistic effect on plant growth (Table **2**).

There is a group of organisms with the least nutrient requirements for survival and hence can flourish in soil that is devoid of any organic matter. Their number varies seasonally and is quite related to water availability in terms of precipitation. These are known as blue-green algae (BGA), including soil species of *Nostoc*, *Anabaena*, *Plectonema*, *Tolypothrix*, and *Scytonema*. Photobioreactors are employed for the mass cultivation of BGA for field applications as they help with nitrogen fixation. Simple shallow trays of iron sheets or bricks or pits lined with polyethylene sheets are also used for mass cultivation.

COMMERCIALLY SIGNIFICANT METABOLITES

The first reported use of microbial metabolites (*e.g.* enzymes) goes back to the year 1915 for converting starch into sugars. Since then, the role of microbial metabolites has been substantially studied, and their applications have reached all the important industries of the present age. These metabolites are directly quoted by genes, which are in turn under stringent regulatory control. Understanding these regulatory controls would help an industrial microbiologist to know the functional feedback control systems existing within the metabolic pathways of the concerned microbe. The end product of each pathway can potentially inhibit the enzymes of the first reaction of the pathway, which is directly linked to the concentration of the end product and its competitive interaction with the allosteric site of the first enzyme of the pathway. This is generally referred to as feedback inhibition, and this knowledge of the pathway can help us modify the control systems *via* mutation, generally targeting the allosteric site of the first enzymes. The first enzyme in the pathway is mutated in such a way that it does not recognize the end product, thus overcoming feedback control systems. This is theoretically fine to say, but the overriding of the feedback control systems alone would not ensure overproduction of end products. It depends on an individual microbial cell to cope with changes, and total cell energy dynamics come into play here. A microbial cell can not, as such, direct all its energy towards overproducing one product of its metabolic pathway. Thus, industrial microbial strains are construed to work in parlance with their modified metabolic systems that give equal chances for a cell to grow, multiply, and produce the desired product in excess. In this regard, *Bacillus* sp. has been found as a new parlance to *Escherichia coli* for the commercial production of heterologous products. *Bacillus*, a common soil bacterium, is playing a larger role in industries due to its abundance of commercially important metabolites, ability to grow faster, and dominance over contaminants in growth media. Deeper genetic studies have thus opened an opportunity to tap the potential of this soil bacterium for industrial applications.

Enzymes

Enzymes operate in a defined cell environment under optimum conditions of neutral pH and mesophilic temperature, and are known to be specific in their action. Microbes have devised a series of enzymes to work for them, both intracellular and extracellular. Most of the enzymes have a predefined biological action that makes cells metabolically efficient. Furthermore, some microbial cells have made enzymes adapted to extreme conditions, be they high or low pH, or psychrophilic or thermophilic temperatures. Some enzymes work at such extremes that they are ideal candidates for industrial applications. Presently, the role of microbial enzymes has increased substantially in the industrial processes, making the overall process environmentally friendly, cheaper, and sustainable with the constant efforts of genetic and protein engineering. Extracellular enzymes are easier to work with as they can be produced in larger concentrations and easily recovered from downstream processing. Intracellular enzymes are produced in limited concentration by the concerned cell, and their extraction at the end of the fermentation process is difficult and tends to be utilized by the cell. Thus, the yield of intracellular enzymes is comparatively lower than extracellular enzymes, thus adding up to their cost of production. The immobilization of enzymes thus helps to utilize their activity to the maximum without losing them physically during the industrial process.

A few enzymes that require cofactors and specific environmental parameters are used with the necessary accessories for better results in the industry. Amylases and glucose isomerase have the widest applications in starch processing industries due to their roles in starch liquefaction and saccharification, along with pullulanases and other endoamylases. Thermostable amylases from *Bacillus licheniformis, B. coagulans*, and *B. amyloliquefaciens* have found applications in the detergent industry. Glucose isomerases produced by *Streptomyces* spp. can catalyze the reversible isomerization of glucose to fructose and are commonly used for the production of alternate sweeteners. Pullulanases from *Bacillus acidopullulyticus* along with cyclodextrinases from *Bacillus coagulans* have industrial applications. Proteases have found a huge role in the detergent, food, and textile industries. Bacterial serine protease, a subtilisin enzyme produced primarily by *Bacillus subtilis*, is widely marketed, and research has shown that these enzymes operate at wide pH and temperature ranges suitable for industrial environments. Glucanases and xylanases are widely used in the brewing industry. Cellulase and lipase are commercially employed in the detergent industry, and especially lipases from *Pseudomonas alcaligenes*, have better adaptability than any other fungal-sourced enzymes. Cellulases have found a role in the textile industry, and xylanases from *Bacillus subtilis* are used in the paper and pulp industry. Cellulase produced by a large number of bacteria and fungi has diversified from its initial applications in bio-ethanol production to the textile industry to improve cotton fabrics. The pharmaceutical industry finds usage of a wide range of microbial enzymes, especially amidases, lipases, and esterases.

Microbial Polymers

Microbial polymers such as dextrans, xanthan, and gellan have found renewed interest and competitive application along with the existing plant and animal-sourced polymers. Some microbial polymers worth mentioning are xanthan produced by *Xanthomonas campestris*, which is used as a foam stabilizer, emulsion stabilizers, and suspension agent in foods; pullulan from *Aureobasidium pullulans* as food coatings; gellan from *Pseudomonas elodea* as a gelling agent in foods; cellulase from *Gluconacetobacter xylinus* finds applications as moisture retention in wound dressings; alginate from *Azotobacter vinelandii* is used as an immobilizing agent in biocatalyst based bioconversion processes in the industry; and curdlans from *Agrobacterium* sp. and *Alcaligenes faecalis* as gelling agents in foods.

Microbial polymers are found to be economical in production, as they sometimes amount to 50% of the dry cell weight. The most significant view of microbial polymers is their role as alternate plastics that are easily biodegradable. Microbes produce these polymers as their outer cell protective coat in the form of exopolysaccharides to protect against drying and keep cells hydrated, helping further in adhesion, biofilm formation, and sometimes pathogenesis. Microbial cellulose is commonly found to be produced by *Acetobacter*, Alcaligenes, *Rhizo-*

bium, and *Sarcina* and is used as a paper additive, thickener, and insulator material.

Amino Acids

Microbes use fermentative and enzymatic methods to produce amino acids, and mostly glycolysis and Kreb's cycle are linked to the production of most of the amino acids. Most of the essential amino acids are tightly regulated by bacterial metabolic feedback control systems. Overcoming both feedback inhibition and feedback repression can lead to an increase in the yield of amino acids. Moreover, the L-amino acids are preferred as they are biologically active. This makes the microbial-based racemic mixture of L and D forms of amino acids more significant than chemically produced. While protein hydrolysis can be the cheaper option for amino acid production, a few very essential amino acids are lacking in plant or animal sources. L-alanine from *Corynebacterium glutamicum* and *Brevibacterium flavum* is used as food additives and has bacteriostatic properties. L- DOPA, the precursor to the neurotransmitter dopamine, is produced by *Erwinia herbicola,* and D-hydroxyphenylglycine (D-HPG), a starting material for semi-synthetic penicillin, is produced by *Pseudomonas, Agrobacterium, Bacillus, Streptomyces,* and *Actinoplanes.*

Solvents

Solvents like butanol, acetone, and isopropanol are fermentative products of commercial value. The fermentative microbes share a few common pathways deviating from pyruvate that lead to the production of acidogenic end products like lactic acid, acetic acid, propionic acid, butyric acid, and solventogenic end products like butanol, acetone, and isopropanol. Yeast like *Saccharomyces* sp. can switch from a respiratory to a fermentative mode that leads to the production of ethanol. Other minor end products of fermentation pathways are carbon dioxide and hydrogen. Sometimes mixed acids are produced. The acetone-butanol fermentation process is historically significant in terms of its role in acetone production during World War I. While butanol is the chief end product of fermentation, acetone, and isopropanol are considered by-products of the process and have additional benefits while extracting the products.

Solvents offer an alternative to petroleum-based products and can be produced on a large scale from agricultural biomass-based feedstocks and can be efficiently converted to biofuel using fermentative microbes. Gram-positive spore-forming bacteria are chiefly known to produce acetone and butanol. Acetone can be produced commercially from *Bacillus macerans* and *Methylosinus trichosporium,* butanol from *Clostridium thermosaccharolyticum.* Butanol and acetone together from C. *acetobutylicum,* C. *pasteurianum,* C. *puniceum,* and C. *saccharobuty-*

licum can be isolated. Isopropanol from *C. acetobutylicum* using molasses as substrate can also be isolated.

Organic Acids

Food grade organic acids of industrial value are sourced *via* fermentative microbes, and some commercially useful metabolites are citric acid, lactic acid, acetic acid, and ascorbic acid. These acids have application as food preservatives as they can inhibit microbial growth and are said to confer aroma on beverages, especially wine. Volatile acids like acetic acid play a significant role in biogas production as they are substrates for methanogenic bacteria for methane production, while propionic acids are inhibitory to the anaerobic digestion process [8]. Industrial production of citric acid employs metabolically improved fungal strains of *Aspergillus*, *Penicillium*, and *Trichoderma*. Acetic acid is industrially produced by *Acetobacter aceti*, *A. pasteurianus*, *Gluconobacter oxydans*, and *Gluconacetobacter liquefaciens*.

Antibiotics

Most soil microbes produce one or more secondary metabolites that are extracellular and targeted against other soil organisms, particularly. This is attributed to the compulsive nature of the soil environment that provides challenging and multi-stress conditions for microbes for their survival. This has resulted in a diverse and potentially beneficial antimicrobial substance that could be isolated from soil microbes. Actinomycetes are known to produce an array of bioactive substances against Gram-positive bacteria [9]. There was a mad scramble to isolate new antibiotics from soil microbes during the early 1960s, driven by reports of the emergence of resistance to penicillin drugs. It was initially produced as it led to the discovery of a range of new antibiotics. During this period, the isolation protocol was standardized, which led to the development of special screening techniques for the isolation of antibiotic producers. But to date, the crowded plate method is the primary screening technique employed for the isolation of antibiotic producers. The mode of action that was deduced for each antibiotic has further led to the grouping of antibiotics into families with similar structures, and this has helped in the specific search for the preferred mode of action and also led to the development of synthetic and semi-synthetic antibiotics.

Secondary metabolites are produced during the stationary phase, which makes the production of antibiotics comparatively easier than primary metabolites. Moreover, these secondary metabolites are extracellular and are produced only during the stationary phase. Historically, these antibiotics have been produced by large scale fermentation. Developments in the fermentation industry in terms of continuous processes and automation of the entire process have made this industry the most sought-after and crucial.

The metabolic engineering of microbes has led to the overproduction of antibiotics by industrial strains, aided by strain improvement with knowledge of the feedback control systems. Secondary metabolites are small molecules that share a common pathway with other primary metabolites. Sometimes, auxotrophic mutants may end up producing an excess of antibiotics as an end product. Beta-lactams are classical antibiotics produced by fungi against many bacterial species. Penicillins and cephalosporins are commercially isolated from *Streptomyces* sp. Aminoglycosides are active against Gram-negative bacteria and are produced by *Streptomyces griseus* under the name of streptomycin, while gentamicins are produced by *Micromonospora purpurea*. Tetracycline and chlortetracycline are isolated from *Streptomyces aureofaciens*, and oxytetracycline from *Streptomyces rimosus*. Antibacterial macrolides like erythromycin are produced by *Saccharopolyspora erythraea*, and polyenes involving amphotericin are produced by *Streptomyces nodosus* with antifungal activity. Glycopeptide antibiotics like vancomycin and teicoplanin are produced by *Amycolatopsis orientalis* and *Actinoplanes teichomyceticus*, respectively. *Streptomyces* alone is known to produce a series of antimicrobial agents with potential applications, which include anthracyclines, avermectin, bafilomycin, renamycin, bisanthraquinone, hygromycin, lincomycin, mitomycin, streptomycin, tetracycline, and sesquiterpene.

BIOMINING

Mining industries employ soil microbes for mineral recovery, which is said to be a low-cost process and is economically profitable. The application of microbes involves extracting metals from low-grade ores that otherwise could not be economically extracted by a normal chemical process. The role of microbes is irreplaceable in terms of their adaptability to extreme pH and high temperatures. For example, copper leaching industries have employed bacteria to mine one-fifth of their total copper production *via* bacterial bioleaching. Most bacteria are employed for the process, to mention a few: *Acidithiobacillus ferrooxidans, A. thiooxidans, Leptospirillum ferrooxidans*, and *L. ferriphilum* [10]. Most of the useful metals exist in metal sulfide form, and examples of a few metals successfully bioleached are copper, iron, gold, uranium, lead, zinc, nickel, and mercury [11 - 13]. Bio leaching is not only economically viable but generates revenue and is environmentally friendly.

MICROBIAL CONSORTIUM

Individual soil microbes are beneficial and possess immense value in terms of primary and secondary metabolites. Sometimes different species of microbes need to come together to achieve a beneficial role. For example, plant growth-promoting bacteria collectively play a significant role in rhizosphere soil. In this regard, instead of looking at microbes individually, it is sometimes more beneficial to view them collectively. Microbes collectively can outperform certain activities; this can be true in the case of bioremediation, wherein a pollutant can be removed from soil by a group of microbes. The consortium also plays a significant role in bioenergy production or decomposition; sometimes decomposition leads to bioenergy production, which is commonly referred to as wealth from waste, *e.g.*, biogas production.

Organic feedstocks like agricultural waste, food waste, and household waste offer a potential substrate for biofuel production, which has industrial relevance [14 - 16]. The present cost escalation in petroleum prices and a bleak future of non-renewable energy resources drying up have created an unstable future for most countries. The economy is tied up with the cost of fuel, and in this sense, soil microbes based on their natural decomposition can produce biofuel, which would be of immense value. Bioethanol has already been vouched as a potential alternate market for utilizable energy that can be produced from energy crops. Now, industries are interested in energy production from organic waste generated from human activities in the form of municipal waste. These organic wastes can be anaerobically digested at a faster rate with the production of methane, a biofuel of immense value that can be used for motor fuel, as an alternative to electricity to light up homes, as manure for organic farming, *etc*. Anaerobic fermentative bacteria include *Bacteroides succinogenes*, *B. cellulosolvens*, *Clostridium cellulovorans*, *C. thermocellum*, and *Syntrophomonas wolfei. Syntrophobacter wolinii*, *Desulfotomaculum thermocisternum*, *Methanosarcina barkeri*, *Methanobacterium formicicum*, *M. thermoautotrophicus, M. bryantii, etc.*, are a few of the known microbes that are found in anaerobic digesters. Microbial consortiums are being developed now that can perform a series of functions, preferably leading to an excess of methane production [17, 18]. Not only does the microbial consortium help in biogas production, but it also makes the entire process stable and sustainable for a longer period of time.

Thus, a consortium of VAM and a host of nitrogen fixers with other plant growth-promoting microbes can be more beneficial and lead to better output in all aspects, whether it be agricultural, environmental, or industrial.

CONCLUDING REMARKS

Soil microbes have wide applications in the field of agriculture as they are primary decomposers that recycle nutrients and are solely responsible for the existing soil matrix. Their ability to fix nitrogen has made them an indispensable tool in agriculture and for improving soil fertility. Environmental aspects of soil microbes like bioremediation, bioleaching, biodegradation, waste recycling, composting, and anaerobic digestion are a few of the benefits mentioned for soil microbes. Diverse industrial aspects have a wide scope in the future for the production of various metabolites of immense commercial value.

CONSENT FOR PUBLICATION

Not applicable.

CONFLICT OF INTEREST

The author declares no conflict of interest, financial or otherwise.

ACKNOWLEDGEMENTS

Declared none.

REFERENCES

[1]   Azizan MS, Zamani AI, Stahmann KP, *et al.* Fungal metabolites and their industrial importance. A brief review. Malaysian J Biochem Mol Biol 2016; 2: 15-23.

[2]   Fukami J, Cerezini P, Hungria M. *Azospirillum*: benefits that go far beyond biological nitrogen fixation. AMB Express 2018; 8: 73.
[http://dx.doi.org/10.1186/s13568-018-0608-1] [PMID: 29728787]

[3]   Harir M, Bendif H, Bellhcene M, *et al. Streptomyces* Secondary Metabolite. Enany S, Ed. Basic Biology and Applications of Actinobacteria. Intech Open 2018; pp. 99-122.

[4]   Jani J, Parvez N, Mehta D. Metabolites of *Pseudomonas*: A New Avenue of Plant Health Management. In: Chakravarthy AK, Ed. New Horizons in Insect Science: Towards Sustainable Pest Management. New Delhi: Springer 2015; pp. 61-9.

[5]   Krishnaraj PU, Shraddha KD. Mineral phosphate solubilizing concepts and prospects in sustainable agriculture. Proc Indian Nat Sci Acad 2014; 80: 389-405.

[6]   Singh P, Ram RM, Singh HB. A Deeper Insight into the Symbiotic Mechanism of *Rhizobium* spp. from the Perspective of Secondary Metabolite. Singh H, Keswani C, Reddy M, *et al.* Eds. Secondary Metabolites of Plant Growth Promoting Rhizomicroorganisms.. Singapore: Springer 2019; pp. 265-91.

[7]   Sivasakthi S, Saranraj P, Sivasakthivelan P. Biological nitrogen fixation by *Azotobacter* sp. – a review. IAJMR 2017; 3: 1274-84.

[8]   Tomei MC, Daugulis AJ. *Ex-situ* bioremediation of contaminated soils: an overview of conventional and innovative technologies. Crit Rev Environ Sci Technol 2013; 43: 2107-39.

[9]   Yogesh BJ, Michael A. Critical factors influencing sustainable operation of anaerobic digesters. Int J Comprehens Res Biol Sci 2015; 2: 32-40.

[10] Glazer AN, Nikaido H, Eds. Microbial Biotechnology: Fundamentals of Applied Microbiology. 2ⁿᵈ ed. Cambridge, UK: Cambridge University Press 2007; pp. 528-9.

[11] Srichandan H, Ranjan KM, Pankaj KP, *et al.* Bioleaching approach for extraction of metal values from secondary solid wastes: a critical review. Hydrometallurgy 2019; 189: 105122.
[http://dx.doi.org/10.1016/j.hydromet.2019.105122]

[12] Mahajan S, Gupta A, Sharma R. Bioleaching and Biomining. In: Singh R, Ed. Principles and Applications of Environmental Biotechnology for a Sustainable Future. Applied Environmental science and Engineering for a sustainable Future. Singapore: Springer 2017; pp. 393-423.

[13] Mishra D, Kim DJ, Ahn JG, *et al.* Bioleaching: a microbial process of metal recovery; a review. Met Mater Int 2005; 11: 249-56.

[14] Bharathi S, Pramod T, Yogesh BJ. Evaluation of urban waste as substrate and its influence on sustainable biomethanation process. Oxford J Sci Res 2018; 1: 28-37.

[15] Yogesh BJ, Michael A. Explicit role of methanogenic consortium developed for improving thermophilic biomethanation of urban waste. Int J Pharma Bio Sci 2013; 4: 1168-77.

[16] Nagamani B, Ramasamy K. Biogas production technology: an Indian perspective. Curr Sci 1999; 77: 44-55.

[17] Batstone DJ, Keller J, Blackall LL. The influence of substrate kinetics on the microbial community structure in granular anaerobic biomass. Water Res 2004; 38: 1390-404.
[PMID: 15016516]

[18] Lee C, Kim J, Hwang K, *et al.* Quantitative analysis of methanogenic community dynamics in three anaerobic batch digesters treating different wastewaters. Water Res 2009; 43: 157-65.
[PMID: 18945471]

CHAPTER 4

Soil Microbes: Role in Agriculture Sustainability

Kishor Chand Kumhar[1] and **Ramesh Nath Gupta**[2,*]

[1] *Deen Dayal Upadhyay Centre of Excellence for Organic Farming, CCS HAU, Hisar – 125004, India*

[2] *Department of Plant Pathology, Bihar Agricultural University, Sabour-813210, India*

Abstract: Soil is the basic and important medium that supports plant and microbial communities for their growth and development. Soil, plants, and microorganisms interact in various ways in nature. The interaction between plants and microbes may be harmful or beneficial in the soil environment. Phytopathogens have harmful effects, whereas antagonists may have beneficial effects on the plant community. Specific Antagonists are capable of controlling phytopathogens through different modes of action. The antagonists may be fungi, bacteria, or actinomycetes under the category of biological control agents (BCAs). Amongst the antagonists, the genus *Trichoderma* is considered a superstar and the most widely exploited biological control agent. Besides plant disease management, it has the potential to enhance vegetative plant growth and resistance against biotic as well as abiotic stresses. In the last couple of years, public interest has been switching from synthetic fungicides to non-chemical fungicides in the agricultural sector. In such a situation, *Trichoderma* spp. could be an ideal option with zero harm to the ecosystem and human health. In India, there are hundreds of manufacturers and marketers of *Trichoderma* products. The majority of its products are available in the form of wettable powder (WP) formulations with variable active ingredients, whereas liquid formulations are very rare. Its formulations are mainly used to manage soil-borne fungal phytopathogens such as species of *Fusarium*, *Pythium*, *Phytophthora*, *Rhizoctonia*, *Verticillium*, *etc.*, of cereals, pulses, vegetables, fruit, and plantation crops. It can also control certain air-borne fungal phytopathogens such as species of *Alternaria*, *Curvularia*, *Colletotrichum*, *etc.* It has great scope in the organic agriculture scenario, and its usage in the crop production system has been increasing day by day. The dose of its application for different crops, diseases, formulations, and manufacturers is variable. However, it should, like synthetic fungicides, be uniform to avoid unnecessary confusion and hesitation among the end-users.

Keywords: Disease management, Phytopathogen, Soil microbes, Stresses, *Trichoderma*, Vegetative growth.

---

* **Corresponding author Ramesh Nath Gupta:** Department of Plant Pathology, Bihar Agricultural University, Sabour-813210, India; E-mail: rameshnathgupta@gmail.com

Ashutosh Gupta, Shampi Jain & Neeraj Verma (Eds.)

INTRODUCTION

Soil is a complex medium as a playground for the survival and development of numerous microorganisms, namely fungi, bacteria, actinomycetes, *etc.* In the soil ecosystem, a rhizosphere is a suitable place where such microbes remain quite active. The rhizosphere provides the required nutrients in the form of root exudates. The root exudates are a rich source of necessary carbohydrates, which attract the microbes of both categories, *i.e.*, harmful and beneficial, and help them grow and develop.

Among the harmful microorganisms, phytopathogens are of immense importance, which causes soil-borne diseases such as pre-and post-emergence seed rot, seedling rot, damping-off, root rots, wilt, collar rot, *etc.* in cereals, vegetables [1], horticultural and plantation crops [2]. Fungal species of *Fusarium*, *Pythium*, *Phytophthora*, *Rhizoctonia*, *Verticillium*, *Sclerotinia*, *etc.*, are the major genera inciting diseases and resulting in huge crop losses.

CHARACTERISTICS OF SOIL-BORNE PHYTOPATHOGENS

The soil-borne phytopathogens cause diseases such as seed and seedling rot, damping-off, stem rot, crown rot, collar rot, and wilt. These pathogens possess certain specific characteristics [3], such as the majority of them being necrotrophic, killing host tissue with the help of their enzymes and toxins. They do not require living cells for their nourishment. They have a wide host range. A single gene can not govern resistance against these pathogens. Due to the unfavourable environmental conditions of the rhizosphere, they keep on surviving in the form of chlamydospore, sclerotia, thick-walled mycelia, and spores. Under favourable conditions, when such forms come into contact with the seed at the time of sowing and the roots of young plants, they reactivate. *Pythium* and *Phytophthora*, fast growing fungi, infect embryos and seeds even before emergence. Most of the pathogens attack young roots. *Fusarium oxysporum* and *Verticillium dahlia* cause wilt disease in plants by entering into vascular tissues, establishing in xylem, and blocking the supply of water and nutrients [4] to the upper parts of the plants. Plants infected with soil-borne pathogens show certain disease symptoms. However, it is very difficult to diagnose properly. The symptoms due to such pathogens resemble stress, drought, and nutrient deficiency.

SIGNIFICANCE OF SOIL-BORNE PHYTOPATHOGENS

The soil-borne phytopathogens are one of the most important decision-makers in successful crop production. Literature reveals that they cause huge crop losses and hence affect the economic status of the growers. Various fungal and bacterial pathogens found in soil have been linked to up to 50% of crop losses [5, 6].

GENUS *TRICHODERMA*

Microorganisms used as biological control agents to protect crops from the attack of insect pests and diseases are of immense importance in agriculture to determine the success of crop production. The biological control agents (BCAs) used for this purpose are naturally found in different kinds of soils to varying extents. This includes the genus *Trichoderma, Beauveria, Metarhizium, Lecanicillium, etc. Trichoderma* spp. is primarily used to manage soil-borne fungal phytopathogens. Deuteromycetous entomopathogenic fungi, namely, *Metarhizium anisopliae, Beauveria bassiana*, and *Paecilomyces fumosoroseus* in soils [7 - 11] can be isolated through proper technique [12]. Numerous factors, such as habitat type, type of crops grown, clay content, pH, and lower organic matter content, affect the occurrence and distribution of entomopathogenic fungi in soil [13, 14].

Beauveria bassiana and *Metarhizium anisopliae* strains were found effective in controlling thrips and whiteflies [15]. *Metarhizium* spp., *Beauveria* spp., and *T. atroviride* showed very good potency against olive fruit fly (*Bactrocera oleae*) and soil-borne *Verticillium dahliae, Phytophthora megasperma*, and *Phytophthora inundata* [16].

Among these beneficial soil-resident fungi, the genus *Trichoderma* has emerged as a superstar as a promising candidate to manage soil-borne fungal diseases [17 - 19]. This chapter is mainly focused on the genus *Trichoderma*, including its various important aspects. *Trichoderma* is commonly found in various soils as a saprophyte on decaying plant materials. It can suppress phytopathogens through competition for space and nutrients. It produces secondary metabolites in the form of antibiotics, which are toxic to phytopathogens. Through hyperparasitism, we can directly manage the phytopathogens in the soil. There are certain genes in the *Trichoderma* species that govern their biocontrol properties [20]. In addition, it encourages vegetative growth and imparts resistance in plants against several phytopathogens and various stresses [21].

Historical Background

Trichoderma spp. as biocontrol agents for the management of phytopathogens were recognized a long time ago. The name *Trichoderma* came into existence in the year 1794 [22]. Later on, its sexual stage as *Hypocrea* sp. was reported [23], which can not be differentiated from *Trichoderma* on a morphological basis. Its systemic position is: **Kingdom:** Fungi, **Phylum:** Ascomycota, **Class:** Euascomycetes, **Order:** Hypocreales, **Family:** Hypocreaceae and **Genus:**

Trichoderma. With the progress of time, its species range has widened, and at present, about 252 species are known. However, of these, only a limited number of species can be utilized as plant protection measures. The economically important species for agricultural application include *T. harzianum, T. viride, T. asperellum, T. atroviride*, and *T. hematum*.

Application of *Trichoderma* in Agriculture

In the current agricultural scenario, *T. harzianum* and *T. viride* are most commonly used for seed, seedling treatment, and soil application to protect cereals, pulses, vegetables, horticultural crops, *etc. T. asperellum* and others are almost negligible in this sector, though *T. asperellum* has the potential to take care of various fungal pathogens. Its wettable powder (WP) and liquid product formulations are available to end users. However, the wettable powder formulations are available in numerous variants. Its foliar spray is beneficial for certain crops to control airborne diseases [24 - 27].

Current Status of *Trichoderma* Formulations

There is a long list of about 970 registered manufacturers with the Central Insecticide Board & Registration Committee (CIB & RC) involved in the manufacturing and marketing of biopesticides in India. Of these, 558 are manufacturing *Trichoderma* spp. (Fig. **1**). It includes private companies, government organizations, and NGOs. Wettable powder (WP) and liquid formulations are available for use by end-users; however, WP formulations in different concentrations (0.5 to 6.0%) are dominant. The liquid formulations contribute less than 2% of total formulations. There are different kinds of formulations [28] and delivery methods to control plant diseases. Onward year 2005, the biopesticides demand has increased tremendously by the year 2016. The compound annual growth rate (CAGR) showed up to 14.1 per cent [29, 30].

Management of Plant Diseases Through Seed Treatment

Seed treatment is an important and economical way to manage the seed-borne phytopathogens [31] at the initial stage with the minimum cost. For this purpose, a small quantity of the desired input is required, and this practise is very easy to handle. For seed treatment, formulations of *T. harzianum, T. viride, etc.*, can be used. Seedlings of vegetable crops can be treated with *T. viride* and *T. harzianum* just before transplanting into the main field (Table **1**).

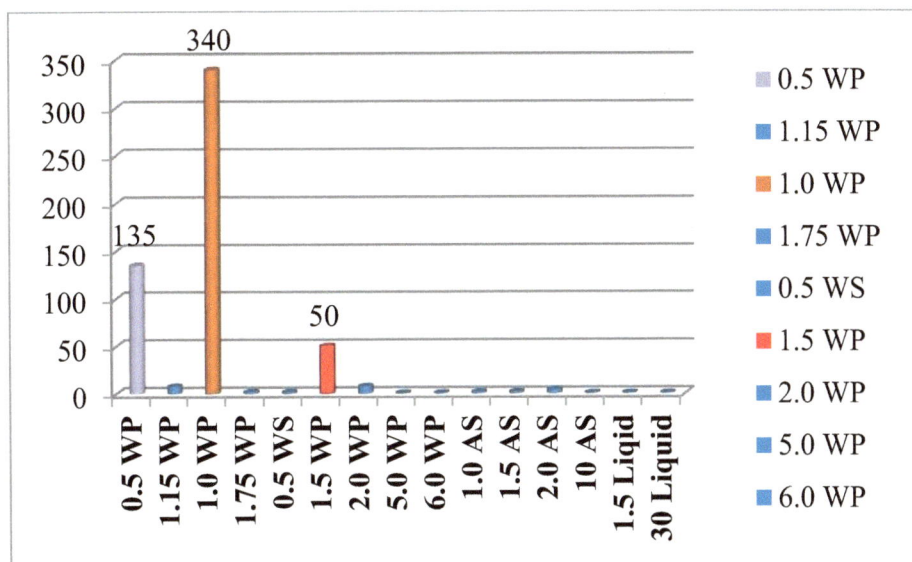

Fig. (1). Number of manufacturers of different *Trichoderma* formulations concentration wise.

Table 1. Seed treatment with promising BCAs for agriculture.

sp./strain of *Trichoderma*	Crop	Adopted by
Trichoderma sp.	Chilli seed treatment	Uttrakhand*
T. viride	Chickpea seed treatment	Madhya Pradesh[#]
T. harzianum	Tomato seedling treatment	Meghalaya[#]
T. viride	Maize seed treatment	Meghalaya[#]
T. viride	Broccoli seed treatment	Sikkim[#]
T. viride	Nursery soil treatment	Sikkim[#]
T. viride	Tomato seed treatment	Sikkim[#]
T. viride	Capsicum seed treatment	Sikkim[#]
T. viride	Chilli seed treatment	Tamil Nadu[#]
T. harzianum	Wheat seed treatment	Uttar Pradesh[#]
Trichoderma sp.	Wheat seed treatment	Uttarakhand[#]
Pseudomonas and *Trichoderma*	Pea seed treatment	Uttarakhand[#]
Pseudomonas and *Trichoderma*	Chickpea seed treatment	Uttarakhand[#]
Trichoderma and *Pseudomonas fluorescence*	Wheat seed treatment	Dharwad[#]
Trichoderma sp.	Potato seed treatment	Dharwad[#]

(Table 1) cont.....

sp./strain of *Trichoderma*	Crop	Adopted by
*National Centre of Organic Farming #Indian Institute of Farming Systems Research		

Management of Plant Diseases Through Soil Treatment

Soil residential phytopathogens such as *Pythium*, *Phytophthora*, *Fusarium*, *Rhizoctonia*, *Sclerotinia*, *Macrophomina*, *etc.* play an important role in the development of various soil-borne diseases such as root rot, wilt, and seedling rot of various crops. Management of such fungal phytopathogens can be achieved through the application of *T. harzianum*, *T. viride*, *etc.*, through soil drenching, BCA-enriched farmyard manure, and vermicompost. *T. harzianum*, *Hypocrea lixii*, *T. atroviride*, *H. atroviridis*, *T. asperellum* and *T. virens* are potential biocontrol agents against phytopathogens [32, 33]. Application of *T. viride*-enriched FYM (5 kg/plant) to two-year guava saplings at the basin near the root zone resulted in decreased wilt incidence and better plant growth in terms of stem girth. *T. harzianum* when applied (50 g/vine) in black pepper field, effectively managed the foot rot disease. For the management of pomegranate wilt, bio-formulation of *T. viride* (0.1 and 0.2%) was found to be significantly superior to the control. *T. viride*, *T. polysporum*, and *T. harzianum* are effective in treating coconut stem bleeding disease [34]. This antagonist reduced the wilt disease of bananas and increased their production significantly when applied to the soil [35]. *T. asperellum* was found inhibitory against *Pythium aphanidermatum*, *P. debaryanum*, *Sclerotium rolfsii*, *Fusarium oxysporum* f. sp. *lycopersici* and *Alternaria solani*. *Trichoderma* (@20 kg/ha) combined with 2.0 tonnes of castor cake/ha reduced the nematode population and increased pomegranate yield. Soil application of silver nanoparticles synthesized from *T. asperellum* resulted in complete control of *Fusarium* wilt in banana cv. Grand Nain (https://icar.org.in/ reports/DARE-ICAR-AR-15-16-english/crop-management-AR-2015-16.pdf). The *T. harzianum* talc formulation could control citrus root rot disease by up to 80% in the lab and up to 65% in the field. (https://icar.org.in/files/DAREAnnual %20Report-2017-18).

Under field conditions, some strains of *Pseudomonas* spp. and *Trichoderma* spp. were found to be effective in controlling banana wilt caused by *F. oxysporum* f. sp. *cubense* (Foc) race 1 [36]. Isolates of *Trichoderma* and *Aspergillus*, when applied in the field for the control of guava wilt disease caused by *F. oxysporum* f. sp. *psidii* and *F. solani*, could reduce disease incidence and promote plant growth. For example, antagonistic fungi like *Aspergillus niger*, *T. harzianum*, and *Penicillium citrinum* have been found effective for the management of the wilt disease of guava [37].

T. asperellum enriched compost made from vegetables and other wastes efficiently reduced tomato wilt caused by *Fusarium oxysporum* f. sp. *lycopersici* [38].

The combination of compost and spent mushroom compost with *T. asperellum*-T34 significantly reduced the severity of wilt disease in carnations and tomatoes. It also suppressed the activity of *Rhizoctonia solani* in cucumber. They antagonistically controlled air-borne fungal pathogens and developed systemic resistance in plants [39].

Trichoderma species have also been found as potential biocontrol agents against the wheat pathogen *Fusarium graminearum* [40].

Management of Plant Diseases Through Foliar Application

Foliar application of BCAs is advisable for the management of airborne fungal and bacterial phytopathogens such as species of *Alternaria, Curvularia, Fusarium, Colletotrichum, Pestalotiopsis, Pyricularia, Puccinia,* and powdery, and downy mildew pathogens of cereal, vegetable, and other crops (Table **2**). The potent microbial candidates include *T. viride, T. harzianum, P. fluorescence,* and *Ampelomyces quisqualis*. Post-prune foliar application of *T. harzianum* and *T. viride* is a common practise in tea (*Camellia* sp.) crop production in Darjeeling and the North East region of India [41]. When banana hands (cv. Grand Nain) were dipped in *T. asperellum* suspension and packed without ethylene absorbent, their shelf life at 13.5°C was extended by up to 75 days. Anthracnose and crown rot were not observed (https://icar.org.in/files/DARE-ICARAnnualReport %202016-17English.pdf). It was noted that *T. viride, T. harzianum,* and *T. asperellum* were potential antagonists for the management of *F. solani* and *Pestalotiopsis theae,* causing dieback and grey leaf spot disease of tea [42]. Foliar spray of *T. asperellum* and *T. atroviride* could manage the dieback disease of tea (*Camellia* sp.) and enhance the vegetative growth in terms of more pluckable shoots [2].

Management of Plant Diseases Through Paste Application

Application of *Trichoderma* paste (20% w/v) is generally done after severe pruning operations (rejuvenation prune and medium prune) in tea plantations to protect the plants from airborne pathogens. Such an application can be useful in horticultural crops in which pruning is done.

Plant Growth Promotion Attributes of *Trichoderma*

It has been mentioned in the literature that *Trichoderma* helps in encouraging the vegetative growth of plants as well as both the yield and quality of produce [43 - 45]. *Trichoderma* produces volatile metabolites such as sesquiterpenes, diterpenes, and tetraterpenes that make plants resistant to pathogens, improving plant health, increasing plant biomass, and chlorophyll content, as well as developing the lateral roots [46].

T. asperellum bioprimed seeds of tomato, brinjal, chilli, okra, ridge gourd, and guar had enhanced seed germination, radicle length, and triggered defense responses against phytopathogens. The *T. asperellum* spore dose depended on the modulation of plant growth in vegetable crops.

WHY IT IS NOT SO POPULAR IN THE FARMING COMMUNITY

It is a naturally occurring and hence safer fungus for human beings, the ecosystem, and the environment as a whole, and hence could be a suitable approach for the management of plant diseases. It has the potential to reduce the agricultural sector's reliance on synthetic pesticides while also conserving natural resources for future use. Currently, synthetic pesticides are dominant in agriculture for certain reasons, such as: they have wide publicity among growers and researchers; they have quick action on targeted pests and pathogens; they are easily available at most locations; and they are easily applicable. In the last couple of years, people have been shifting towards safe and good agricultural practises with the alternative of synthetic pesticides.

Table 2. Foliar spray of biological control agents in agriculture.

sp./strain of *Trichoderma*	Crop disease	Adopted by
Trichoderma sp.	Anthracnose of chilli	Uttrakhand*
Trichoderma sp.	Rust, powdery mildew, and blight of pea	Uttarakhand[#]
Trichoderma sp.	Wilt and blight of chickpea	Uttarakhand[#]
Trichoderma harzianum	Black rust, brown rust, yellow rust, and leaf blight of wheat	Jharkhand[#]
T. harzianum or *T. viride*	Loose smut of wheat	Jharkhand[#]
*National Centre of Organic Farming [#]Indian Institute of Farming Systems Research		

LIMITATIONS OF BIOPESTICIDES

Registration

The registration procedure is a bit complicated and time-consuming, taking around 3-4 years in India. There are so many formalities, such as data generation of bioefficacy, toxicological studies, *etc.* Likewise, there are certain regulations in the European Union that adhere to the registration of biopesticides [47].

Variable Dose of Application

Globally, it is unanimously accepted that antagonist *Trichoderma* spp. is quite efficient in taking care of numerous soil, seed, and airborne fungal phytopathogens of crops in the agriculture sector. However, like synthetic pesticides, for example, copper oxychloride 50% wettable powder (WP) is used at a rate of 500 g per 200 litres of water, meaning 2.5 g per liter of water. The question here is whether each manufacturer uses the same active ingredient and concentration.

However, in the case of *Trichoderma* spp. formulation, this thing varies from manufacturer to manufacturer. If there were manufacturers A, B, C, and so on, their product active ingredient as well as application dose in a particular segment, say seed treatment, would be variable. Some advocate 2 g, some 4 g, some 5 g, and so on. So it should be unified, say 5 g per kg seed, so that end users may use it uniformly. Based on the active ingredients, the application doses for seedbed treatment, drenching, and foliar spray should be unified.

FACTORS AFFECTING INTERACTION OF PATHOGENIC AND BENEFICIAL FUNGI IN SOIL

There are so many factors that affect the interaction of both pathogenic and beneficial fungi in the soil [3]. The important ones are soil tillage, organic amendments through the application of farmyard manure, compost, vermicompost, green manuring, *etc.*

CONCLUDING REMARKS

Soil microbes play a vital role in sustaining crop production in various ways. The soil microbes regulate different soil reactions to maintain ideal conditions in favour of plant growth and development, which ultimately reflects in increased crop yield. There are certain soil-inhabitant beneficial fungi, better known as antagonists, which are efficient in controlling the soil-borne fungal phytopathogens that cause diseases such as seed rot, damping-off, root rot, and wilt. *Trichoderma* spp. are the promising representative candidates for antagonists. These are naturally available in different habitats variably and can be developed in the form of product formulations by various methods to be used in agriculture for seed treatment and so many other purposes to take care of plant diseases. Hundreds of manufacturers in India and abroad are active in producing its various usable formulations. Such formulations have great scope in agriculture, particularly "organic agriculture", as far as current and future agricultural perspectives are concerned. Such products are much safer for the associated environment, being natural in origin and free from health hazards. In the current agricultural scenario, such bioagents should be popularized on a large scale to create a healthy environment. Product quality is an important issue that determines its performance under field conditions and should be taken into account. Secondly, the dose of application for various purposes seems quite variable among the products manufactured by various manufacturers that create confusion. Therefore, it should be unified like other agrochemicals.

CONSENT FOR PUBLICATION

Not applicable.

CONFLICT OF INTEREST

The author declares no conflict of interest, financial or otherwise.

ACKNOWLEDGEMENTS

Authors are grateful to Dr. Azariah Babu, Principal Scientist cum Deputy Director (Research) for his valuable suggestions and critically checking the chapter.

REFERENCES

[1]   Punja ZK, Utkhede RS. Using fungi and yeasts to manage vegetable crop diseases. Trends Biotechnol 2003; 21: 400-7.
[PMID: 12948673]

[2]   Kumhar KC, Babu A. Economically Important Diseases of Tea (*Camellia* sp.) and Their Management. Chand G, Akhtar MN, Kumar S, Eds. Diseases of Fruits and Vegetable Crops: Recent Management Approaches. 1st ed. Boca Raton: Apple Academic Press 2020; pp. 435-59.

[3]   Raaijmakers JM, Paulitz TC, Steinberg C, *et al.* The rhizosphere: a playground and battlefield for soil-borne pathogens and beneficial microorganisms. Plant Soil 2009; 321: 341-61.

[4]   Ortiz E, Cruz M, Melgarejo LM, *et al.* Características hispatologicas da infecção causada por *Fusarium oxysporum* e *F. solani* em maracujá-roxo (*Passiflora edulis* Sims). Summa Phytopathol 2014; 40: 134-40.

[5]   Oerke EC, Dehne HW. Safeguarding production-losses in major crops and the role of crop protection. Crop Prot 2004; 23: 275-85.

[6]   Lewis JA, Papavizas GC. Biocontrol of plant diseases: the approach for tomorrow. Crop Prot 1991; 10: 95-105.

[7]   Sookar P, Bhagwant S, Ouna EA. Isolation of entomopathogenic fungi from the soil and their pathogenicity to two fruit fly species (Diptera: Tephritidae). J Appl Entomol 2008; 132: 778-88.

[8]   Bing LA, Lewis LC. Occurrence of the entomopathogen *Beauveria bassiana* (Balsamo) Vuillemin in different tillage regimes and in *Zea mays* L. and virulence towards *Ostrinia nubilalis* (Hübner). Agric Ecosyst Environ 1993; 45: 147-56.

[9]   Doberski JW, Tribe HT. Isolation of entomogenous fungi from elm bark and soil with reference to ecology of *Beauveria bassiana* and *Metarhizium anisopliae*. Trans Br Mycol Soc 1980; 74: 95-100.

[10]   Meyling NV, Eilenberg J. Occurrence and distribution of soil-borne entomopathogenic fungi within a single organic agroecosystem. Agric Ecosyst Environ 2006; 113: 336-41.

[11]   Sevim A, Demir I, Höfte M, *et al.* Isolation and characterization of entomopathogenic fungi from hazelnut-growing region of Turkey. BioControl 2010; 55: 279-97.

[12]   Meyling NV. Methods for isolation of entomopathogenic fungi from the soil environment. Manual for Isolation of Soil-borne Entomopathogenic Fungi, Laboratory Manual 2007; pp. 1-18.

[13]   Quesada-Moraga E, Navas-Cortés JA, Maranhao EAA, *et al.* Factors affecting the occurrence and distribution of entomopathogenic fungi in natural and cultivated soils. Mycol Res 2007; 111: 947-66.
[PMID: 17766099]

[14]   Bidochka MJ, Kasperski JE, Wild GAM. Occurrence of the entomopathogenic fungi *Metarhizium anisopliae* and *Beauveria bassiana* in soils from temperate and near-northern habitats. Can J Bot 1998; 76: 1198-204.

[15]   Sánchez-Peña SR, Lara JSJ, Medina RF. Occurrence of entomopathogenic fungi from agricultural and natural ecosystems in Saltillo, México, and their virulence towards thrips and whiteflies. J Insect Sci 2011; 11: 1.
[http://dx.doi.org/10.1673/031.011.0101] [PMID: 21521145]

[16]   Lozano-Tovar MD, Ortiz-Urquiza A, Garrido-Jurado I, *et al.* Assessment of entomopathogenic fungi and their extracts against a soil-dwelling pest and soil-borne pathogens of olive. Biol Control 2013; 67: 409-20.

[17]   Kumar G, Maharshi A, Patel J, *et al. Trichoderma* : A potential fungal antagonist to control plant diseases. SATSA Mukhapatra - Ann Tech Issue 2017; 21: 206-18.

[18]   Sanjeev K. *Trichoderma*: a biological weapon for managing plant diseases and promoting sustainability. Int J Agric Sci Vet Med 2013; 1: 106-21.

[19]   Naher L, Yusuf UK, Ismail A, *et al. Trichoderma* spp.: a biocontrol agent for sustainable management of plant diseases. Pak J Bot 2014; 46: 1489-93.

[20]   Kubicek CP, Mach RL, Peterbauer CK, *et al. Trichoderma*: from genes to biocontrol. J Plant Pathol 2001; 83: 11-23.

[21]   Shoresh M, Harman GE. The molecular basis of shoot responses of maize seedlings to *Trichoderma harzianum* T22 inoculation of the root: a proteomic approach. Plant Physiol 2008; 147: 2147-63.
[PMID: 18562766]

[22]   Persoon CH. Dispositio methodica fungorum. Neues Mag für die Bot 1794; 1: 81-128.

[23]   Tulasne LR, Tulasne C, Eds. Selecta Fungorum Carpologia: Nectriei-Phacidiei-Pezizei. Parisiis, Imperatoris Jussu, In Imperiali Typographeo Excudebatur 1865; p. 108.

[24]   Lo CT, Nelson EB, Harman GE. Improved biocontrol efficacy of *Trichoderma harzianum* 1295-22 for foliar phases of turf diseases by use of spray applications. Plant Dis 1997; 81: 1132-8.
[PMID: 30861707]

[25]   Navaneetha T, Prasad RD, Venkateswar RL. Liquid formulation of *Trichoderma* species for management of gray mold in castor (*Ricinus communis* L.) and *Alternaria* leaf blight in sunflower (*Helianthus annuus* L.). J Biofert Biopest 2015; 06: 1.
[http://dx.doi.org/10.4172/2155-6202.1000149]

[26]   Dal Belloa G, Rollána MC, Lampugnani G, *et al.* Biological control of leaf grey mould of greenhouse tomatoes caused by *Botrytis cinerea.* Int J Pest Manage 2011; 57: 177-82.

[27]   Abdel-Kader MM, El-Mougy NS, Aly MDE, *et al.* Integration of biological and fungicidal alternatives for controlling foliar diseases of vegetables under greenhouse conditions. Int J Agric For 2012; 2: 38-48.

[28]   Woo SL, Ruocco M, Vinale F, *et al. Trichoderma*-based products and their widespread use in

agriculture. Open Mycol J 2014; 8: 71-126.

[29]   Lehr P. Global markets for biopesticides. BCC Res CHM029E. 2014; p. 177.

[30]   Thakore Y. The biopesticide market for global agricultural use. Ind Biotechnol (New Rochelle NY) 2006; 2: 194-208.

[31]   Coşkuntuna A, Özer N. Biological control of onion basal rot disease using *Trichoderma harzianum* and induction of antifungal compounds in onion set following seed treatment. Crop Prot 2008; 27: 330-6.

[32]   Jeger MJ, Jeffries P, Elad Y, *et al.* A generic theoretical model for biological control of foliar plant diseases. J Theor Biol 2009; 256: 201-14.
[PMID: 18983855]

[33]   Hjeljord L, Tronsmo A. *Trichoderma* and *Gliocladium* in biological control: an overview. Harman GE, Kubicek CP, Eds. *Trichoderma* and *Gliocladium*. Enzymes, Biological Control and Commercial Applications. UK: Taylor and Francis London 1998; Vol. 2: pp. 129-55.

[34]   Kannangara S, Dharmarathna RMGCS, Jayarathna DL. Isolation, identification and characterization of *Trichoderma* species as a potential biocontrol agent against *Ceratocystis paradoxa.* J Agric Sci 2017; 12: 51-62.

[35]   Thangavelu R, Gopi M. Combined application of native *Trichoderma* isolates possessing multiple functions for the control of *Fusarium* wilt disease in banana cv. Grand Naine. Biocontrol Sci Technol 2015; 25: 1147-64.

[36]   Bubici G, Kaushal M, Prigigallo MI, *et al.* Biological control agents against *Fusarium* wilt of banana. Front Microbiol 2019; 10: 616.
[http://dx.doi.org/10.3389/fmicb.2019.00616] [PMID: 31024469]

[37]   Mishra AK. Present status of important diseases of guava in India with special reference to wilt. Acta Hortic 2007; 1013: 507-23.

[38]   Cotxarrera L, Trillas-Gay MI, Steinberg C, *et al.* Use of sewage sludge compost and *Trichoderma asperellum* isolates to suppress *Fusarium* wilt of tomato. Soil Biol Biochem 2002; 34: 467-76.

[39]   Segarra G, Sant D, Trillas MI, *et al.* Efficacy of the microbial control agent *Trichoderma asperellum* strain t34 amended to different growth media against soil and plant leaf pathogens. Acta Hortic 2013; 1013: 515-20.

[40]   Hagn A, Pritsch K, Schloter M, *et al.* Fungal diversity in agricultural soil under different farming management systems, with special reference to biocontrol strains of *Trichoderma* spp. Biol Fertil Soils 2003; 38: 236-44.

[41]   Barthakur BK, Dutta P. Disease Management in Tea. In: Goswami BK, Ed. Tea Field Management. Tocklai Experimental Station, Jorhat, Assam: India: Tea Research Association 2011; pp. 182-8.

[42]   Kumhar KC, Babu A, Bordoloi M, *et al.* Comparative bioefficacy of fungicides and *Trichoderma* spp. against *Pestalotiopsis theae*, causing grey blight in tea (*Camellia* sp.): an *in vitro* study. Int J Curr Res Biosci Plant Biol 2016; 3: 20-7.

[43]   Berg G. Plant-microbe interactions promoting plant growth and health: perspectives for controlled use of microorganisms in agriculture. Appl Microbiol Biotechnol 2009; 84: 11-8.
[PMID: 19568745]

[44]   Stewart A, Hill R. Applications of *Trichoderma* in Plant Growth Promotion. In: Gupta VK, Schmoll M, Herrera-Estrella A, Eds. *et al.* Biotechnology and Biology of Trichoderma. USA: Elsevier BV, MA 2014; pp. 415-28.

[45]   Pascale A, Vinale F, Manganiello G, *et al.* *Trichoderma* and its secondary metabolites improve yield and quality of grapes. Crop Prot 2017; 92: 176-81.

[46]   Lee S, Yap M, Behringer G, Hung R, Bennett JW. Volatile organic compounds emitted by *Trichoderma* species mediate plant growth. Fungal Biol Biotechnol 2016; 3: 7.
[http://dx.doi.org/10.1186/s40694-016-0025-7] [PMID: 28955466]

[47]   Zaki O, Weekers F, Thonart P, *et al.* Limiting factors of mycopesticide development. Biol Control 2020; 144: 104220.
[http://dx.doi.org/10.1016/j.biocontrol.2020.104220]

CHAPTER 5

Climate Change and its Influence on Soil Microbial Community

Jitendra Kumar[1,*], Nishant K. Sinha[1], M. Mohanty[1], Alka Rani[1], R.S. Chaudhary[1] and Avinash Pandey[2]

[1] *ICAR-Indian Institute of Soil Science, Nabibagh, Berasia Road, Bhopal, India*

[2] *ICAR-Indian Institute of Agricultural Biotechnology, Ranchi, 834010, India*

Abstract: The effects of climate change on crop yields vary greatly from region to region across the globe. The projected climate change will also adversely affect soil quality by changing its physiochemical and biological properties. The soil's biological properties and processes are primarily mediated by microbial diversity and their distribution. The presence of soil microbes facilitates the production of greenhouse gases (GHGs). The microorganism also responded to global warming and climate change by either producing greenhouse gases or utilizing them in the environment. Soil microorganisms can recycle and transform the essential elements such as carbon and nitrogen that make up cells. Even small changes in the soil moisture content result in a change in the microbial habitat, particularly the fungal communities. However, the bacterial communities remain intact. The increase in the concentration of greenhouse gases like carbon dioxide not only increases methane production from the soil but also reduces the uptake of methane by up to 30% in the soil microbial population. The microbial communities of the tree leaves act on plant residue during this process. The increase in temperature is likely to accelerate the rate of decomposition that emits carbon dioxide from the soil. However, higher temperatures also elevate soil nitrogen levels, which suppresses the rates of fungal decomposition. This affects microbial communities. At the same time, trees and shrubs that advance towards the north in the tundra under the influence of temperature alteration can also influence microbes in unknown ways through the shadows they cast on the ground.

Keywords: Climate change, Greenhouse gases, Soil health, Soil microbial communities.

INTRODUCTION

Agriculture is facing several challenges in the 21st century, and the most notable is climate change. The human population is also expected to reach 9.7 billion by 2050, which will further pressurize world agriculture.

---

* **Corresponding author Jitendra Kumar:** ICAR-Indian Institute of Soil Science, Nabibagh, Berasia Road, Bhopal, India; E-mail: jitendra.iari@gmail.com

Ashutosh Gupta, Shampi Jain & Neeraj Verma (Eds.)

Climate change is posing pressure on humanity's ability to feed itself. The effects of climate change on crop yields vary greatly from region to region across the globe. Recent IPCC reports (https://www.ipcc.ch/srccl/chapter/chapter-5/.) show that climate change negatively affects crop yields (*e.g.*, maize and wheat) in many lower-latitude regions; however, yields of some crops (*e.g.*, maize, wheat, and sugar beets) have been positively affected in many higher-latitude parts. Warming compounded by drying has caused large adverse effects on yields in parts of the Mediterranean. Based on indigenous and local knowledge (ILK), climate change is affecting food security in the drylands, particularly those in Africa and the high mountain regions of Asia and South America. The projected climate change will also adversely affect soil health by changing the physical, chemical, and biological parameters of the soil. The biological parameters of soil and processes are primarily associated with microbial diversity and their population. The presence of soil microbes facilitates making and using greenhouse gases. The microorganism also responded to global warming and climate change by either generating greenhouse gases (GHGs) or utilizing these gases in the environment. Soil microorganisms can recycle and transform the essential elements such as carbon and nitrogen that make up cells. In this chapter, an effort is made to describe the response and behaviour of soil microbes under climate change scenarios.

CLIMATE CHANGE

Climate is defined as the long-term average of weather variables, including temperature, rainfall, humidity, and wind in a particular region. The climate is a complex, interactive system consisting of the atmosphere, land surface, snow and ice, oceans, other water bodies and living things. The climate of any location is a function of latitude, longitude, earth's axial tilt, earth's wind belts, the difference in temperatures between land and sea, and topography. The earth is surrounded by a thick layer of gases that keeps the planet warm and allows plants, animals, and microbes to live. These gases work like a blanket. Without this blanket, the earth would be 20–30°C colder and much less suitable for life. However, anthropogenic production of greenhouse gases promotes the absorption of outgoing longwave radiation, leading to increased temperatures and causing climate change and global warming. The greenhouse effect (GFE) is the phenomenon whereby the earth's atmosphere traps solar radiation and is mediated by gases such as carbon dioxide (CO_2), water vapor, and methane (CH_4) that allow incoming sunlight to pass through but absorb the heat radiated back from the earth's surface. It provides a blanketing effect in the lower strata of earth's atmosphere, and this blanketing effect is being enhanced by human activities like the burning of fossil fuels.

Causes of Climate Change

The earth's climate is affected by natural factors (Fig. **1**) such as solar output, the heat generated through volcanic eruptions, and other natural phenomena that primarily result in a change in the amount of incoming radiation. These changes in incoming solar irradiance have had a significant effect on contributing to climate trends over the last century. This is further aggravated due to the industrial revolution. The effect of GHG emissions on the atmosphere has been over fifty times greater than that of changes in the sun's output.

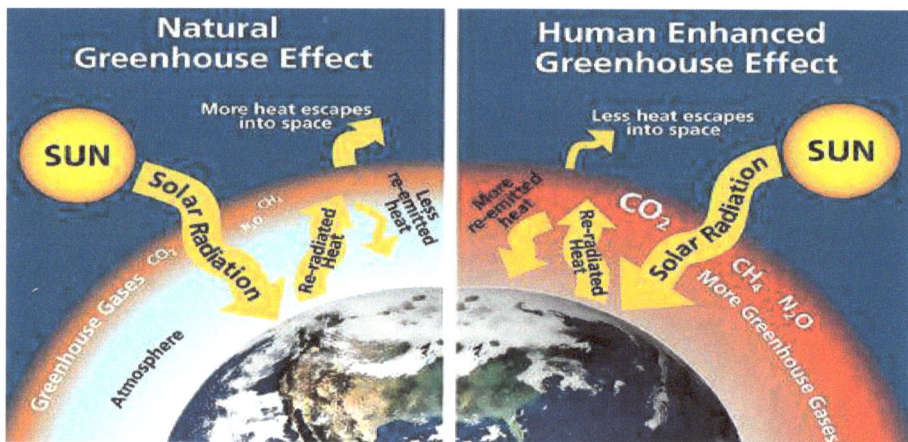

Fig. (1). Causes of climate change, natural and anthropogenic. (Source: https://www.climatetheory.net).

Apart from these natural causes, anthropogenic activities (Fig. **2**), such as the burning of fossil fuels like petroleum and coal, and the utilization of forest land for human settlement and agricultural activities, are the main drivers of climate change. These human activities have increased substantially since the post-industrial revolution, influencing the climate system. These activities change the soil surface temperature and emit various substances, apart from the GHGs, into the atmosphere. These can alter both the amount of energy received by the earth's surface and the amount of energy emitted, which can have both warming and cooling effects on the earth's surface. The main product of fossil fuel burning is carbon dioxide, which is a greenhouse gas. Enhanced emissions of other greenhouse gases and their natural build-up in the atmosphere have led to an enhancement of the natural greenhouse effect. However, human-induced enhancement of the concentration of greenhouse gases traps the outgoing emissions and, as a result, increases the temperature of the earth. Such a shift in earth's temperature could have far-reaching and unpredictable environmental, social, and economic ramifications.

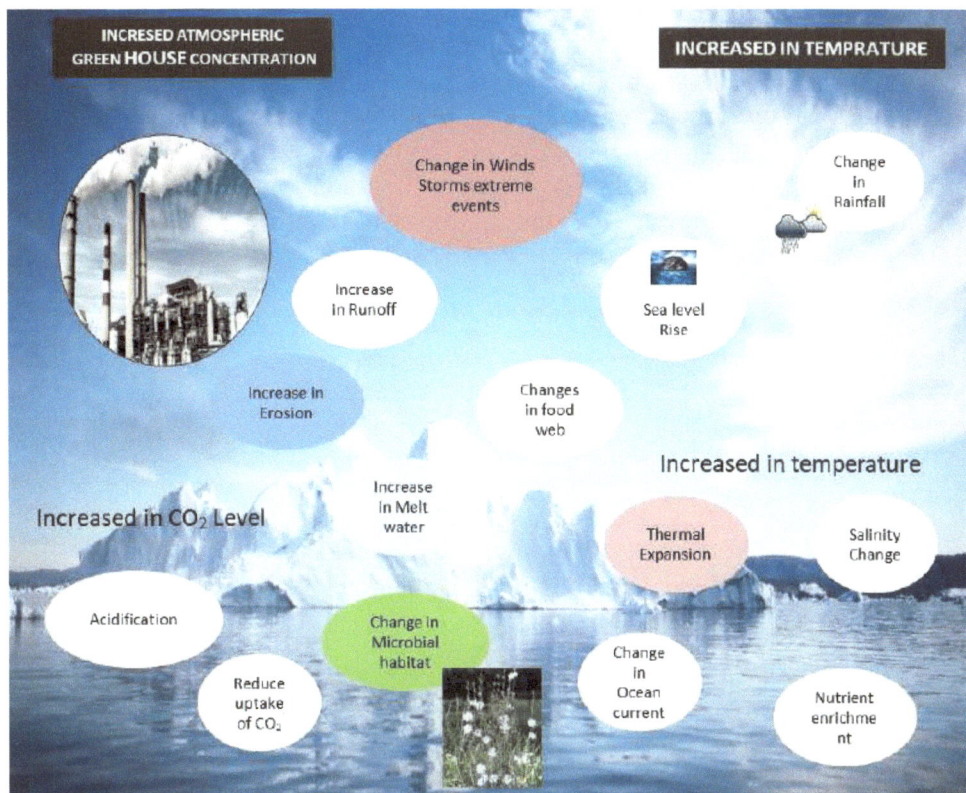

Fig. (2). The impact of climate change on different spheres of life.

Impact of Climate Change

Ecosystems are also affected by climate change. Habitats are being modified by the timing of events such as flowering and egg-laying. The natural habitats of soil microbes are shifting, and species are altering their home ranges. Changes are also witnessed in the ocean. The ocean absorbs approximately 30% of the total carbon dioxide released to the atmosphere from the burning of fossil fuels, which results in acidification of the ocean, which is becoming more acidic and affecting marine life. The elevated temperature causing thermal expansion and melting of the land ice sheets and glaciers puts coastal areas at greater risk of inundation, erosion, and storm surge.

SOIL MICROBIAL COMMUNITY

Microbes play an important role in every aspect of life on Earth due to their extensive diversity in form and function. One teaspoon of topsoil contains around 1 billion individual microscopic cells and around 10,000 different species in soil.

Several microbes (Table **1**) in the soil exist, but they possess less biomass due to their small size. Actinomycetes are a factor of 10 times smaller in number but are larger, so they are similar in biomass to bacteria; the fungal population is smaller in number, but they dominate the soil biomass when the soil is not disturbed. Bacteria, actinomycetes, and protozoa are hardy. They can tolerate more soil disturbances than fungal populations. Hence, they dominate in tilled soils, while fungal and nematode communities tend to dominate in untilled soils with about 80 to 150 quintals of bacteria, fungi, protozoans, nematodes, earthworms, and arthropods.

Table 1. Different soil microbial communities with their biomass.

S. No.	Microorganisms	Number/g of soil	Biomass (g/m²)
1	Bacteria	10^8–10^9	40–500
2	Actinomycetes	10^7–10^8	40–500
3	Fungi	10^5–10^6	100–1500
4	Algae	10^4–10^5	1–50
5	Protozoa	10^3–10^4	Varies
6	Nematodes	10^2–10^3	Varies

(Source: https://ohioline.osu.edu/factsheet/SAG-16)

ROLES OF MICROBES IN SOIL ECOSYSTEMS

The soil microbial community plays a vital role in the biogeochemical cycle in the soil. The microbes ensure the turnover and supply of nutrients essential for plant and crop growth through the interconversion of different forms of nitrogen, sulfur, and phosphorus, which are interlinked with the carbon cycle. Microorganisms are responsible for the degradation of organic matter, which not only controls the release of plant nutrients into the soil, but also acts for the maintenance of soil structure and sustainability of soil quality for plant growth. Microorganisms are required for remediation, through degradation of organic pollutants and immobilization of heavy metals, providing obvious examples of improving soil quality. Microbial activity in soil is also responsible for carbon losses to the atmosphere through respiration and methanogenesis. They purify the environment, and mostly water, through the degradation of pollutants (Fig. **3**).

DRIVERS OF THE MICROBIAL COMMUNITIES IN THE SOIL

The spread of microbial communities depends on edaphic properties, rainfall, and temperature that greatly contribute to their ability to grow and multiply at large ecological scales [1, 2]. It is reflected in the dominance of acid bacteria and the alphaproteobacteria in acidic soil. The strong influence of pH on bacterial community composition has also been identified for ectomycorrhizal fungi (EMC) inhabiting forest soils [3, 4]. However, pH appears to be the primary driver of bacterial community composition. Other physicochemical and climatic parameters have also significantly influenced the survival and activities of bacteria. In the case of fungi, precipitation has a strong effect on community richness. The majority of fungal taxonomic and functional groups have distinct preferences for specific edaphic conditions such as pH, calcium, and phosphorus soil content [5].

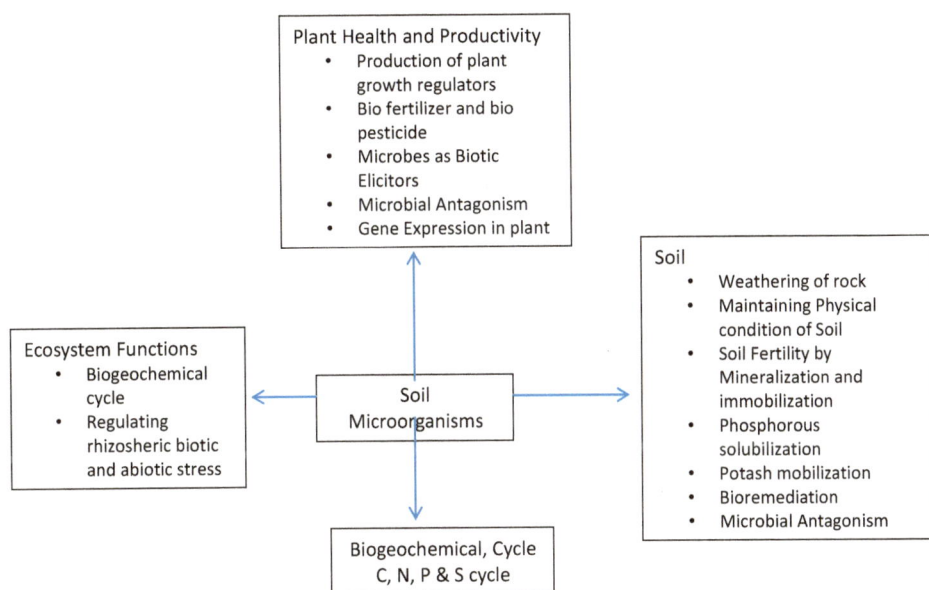

Plant Health and Productivity
- Production of plant growth regulators
- Bio fertilizer and bio pesticide
- Microbes as Biotic Elicitors
- Microbial Antagonism
- Gene Expression in plant

Soil
- Weathering of rock
- Maintaining Physical condition of Soil
- Soil Fertility by Mineralization and immobilization
- Phosphorous solubilization
- Potash mobilization
- Bioremediation
- Microbial Antagonism

Ecosystem Functions
- Biogeochemical cycle
- Regulating rhizosheric biotic and abiotic stress

Soil Microorganisms

Biogeochemical, Cycle C, N, P & S cycle

Fig. (3). Schematic diagram of the roles of microbial communities in soil and plant ecosystems.

The studies also reported that soil microbial biomass is primarily driven by moisture and soil nitrogen content, while temperature has an indirect effect at global scales on microbial biomass (Table **2**). The studies also reported that soil microbial biomass is primarily driven by moisture and soil nitrogen content, while temperature has an indirect effect at global scales on microbial biomass (Table **2**). The microbial world encompasses most of the phylogenetic diversity on Earth, like all bacteria, all archaea, and most lineages of the eukarya, which are microorganisms. Microbes live in every kind of habitat (terrestrial, aquatic, atmospheric, or living host), and their presence invariably affects the environment in which they grow. Their diversity enables them to thrive in extremely cold or hot climates. Their diversity also makes them tolerant of many other conditions, such as limited water availability, high salt content, and low oxygen levels. The response of soil microbial communities to climate change is unknown as to how interactions between plants and soil microbial communities act in response to climate change. They will probably affect the outcome of plant-soil feedback, with possible consequences for vegetation dynamics, ecosystem processes, and feedback effects on climate. For instance, drought can change the population dynamics of soil bacterial and fungal communities and change the interaction between soil-plants, and microbes through feedback and plant competitive interactions [6]. Soil bacterial populations in the tropics and high latitudes are subjected to higher levels of environmental filtering, potentially making these communities more vulnerable to global climate change [7]. As such, climate change potentially has both strong direct effects on plants and soil microorganisms and indirect effects through altered plant metabolism and the quality and quantity of resources entering into the soil. This will likely have great consequences for plant-soil feedbacks, a system of vegetation, and feedback effects on local or even global climatic conditions (Fig. **4**).

Table 2. Assessment of potential impacts of climate change drivers on soil microbial community [8].

Climate Change Drivers					
Soil microbes	Warming	Drier conditions	Wetter conditions	Fire	Increased CO_2
Fungi					
Pathogen	+++	−	+++	−	0
Saprotrophic/organic matter decomposers	++	−	+++	−	?
Arbuscular mycorrhiza	+++	+++	+	−	?
Ectomycorrhiza	+	−	+	−	+
Bacteria					
Pathogens	+++	−	+++	−	?
Symbiotic N_2 fixers	0	−	++	0	+
Nonsymbiotic N_2 fixers	0	−	++	0	+
Drought-tolerant microbes	+	+++	−	0	?
Other specific coevolved microbes	+++	−	++	++	?
Others					
Primary detritivorous invertebrates	++	−	++	−	?
Secondary detritivorous invertebrates	++	−	++	−	?
Root herbivores	++	−	++	0	?

```
                    ┌─────────────────────────────────┐
                    │  Drivers of Microbial Community  │
                    └─────────────────────────────────┘


       ┌──────────────────────┐            ┌──────────────────────────┐
       │  Local Level Drivers │            │ Continental Level Drivers│
       └──────────────────────┘            └──────────────────────────┘
```

Soil Factors	Crop physicochemical Factors	
• pH		
• Organic matter		• Precipitation
• Moisture	• Litter quality	• Temperature
• Nutrient content	• Root traits	• Edaphic conditions
• Cations	• Root exudation	

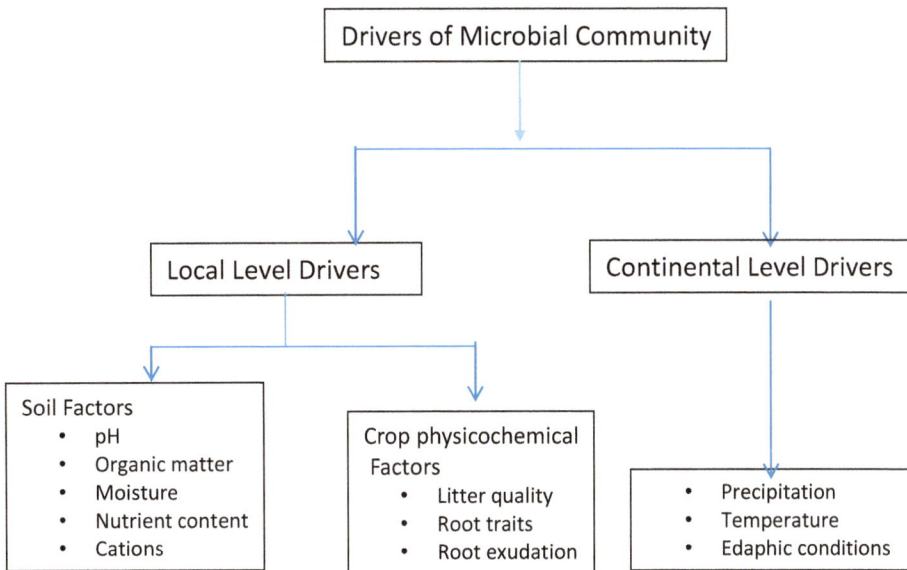

Fig. (4). Various drivers of the microbial community in soil.

The dynamics operating in the plant modulation of soil carbon sequestration are complex and involve numerous biotic interactions between plants, their symbionts (*i.e.*, nitrogen-fixing bacteria and mycorrhizal fungi, and decomposer microorganisms). The rate of organic matter decomposition is determined by microbial activity; thus, the greater the activity of the microbes, the faster the rate of decomposition and thus the loss of carbon from the soil. Keeping this in view, De Deyn and coworkers [9] developed a simplified plant behavior-based substructure for a comprehensive understanding of the relationship between plant communities, the microbial community under climate influence, and soil carbon sequestration (Fig. **5**). It is in line with the growing roles of plant attributes in soil nutrient and carbon cycling. Certain plant attributes can select for particular groups of soil microbes that play vital roles in these processes of nutrient cycling [10]. For instance, a recent study showed that nitrogen and carbon in the root were in synchronization with soil properties, particularly the biomass and concentration of nitrogen. The presence of low plant biomass slowed down the rate of decomposition of recalcitrant soil organic matter, whereas increasing plant biomass enhanced rates of decay [11]. However, specific plant attributes such as differences in the number of root exudates and their chemical composition are also likely to affect soil carbon's mineralization by changing microbial activity [11]. These studies indicate that the plant attributes, especially those of root biomass, provide a potential tool for a project. The directional changes in plant species distribution with global climate change will affect the Earth [10]. However, given the importance of root attributes for soil microbial carbon cycling, modification of root traits of plants, including the development of perennial grains with more root mass and root depth, may provide a way to enhance soil carbon sequestration while at the same time producing food grain [12, 13]. Indeed, in a recent study, the authors explicitly tested the phylogenetic clustering of the responses of fungal and bacterial community compositions to both nitrogen addition and drought and found that the fungal community response to drought had greater phylogenetic clustering than the response to nitrogen addition, while the phylogenetic clustering of bacterial community responses was equal for these disturbances [14].

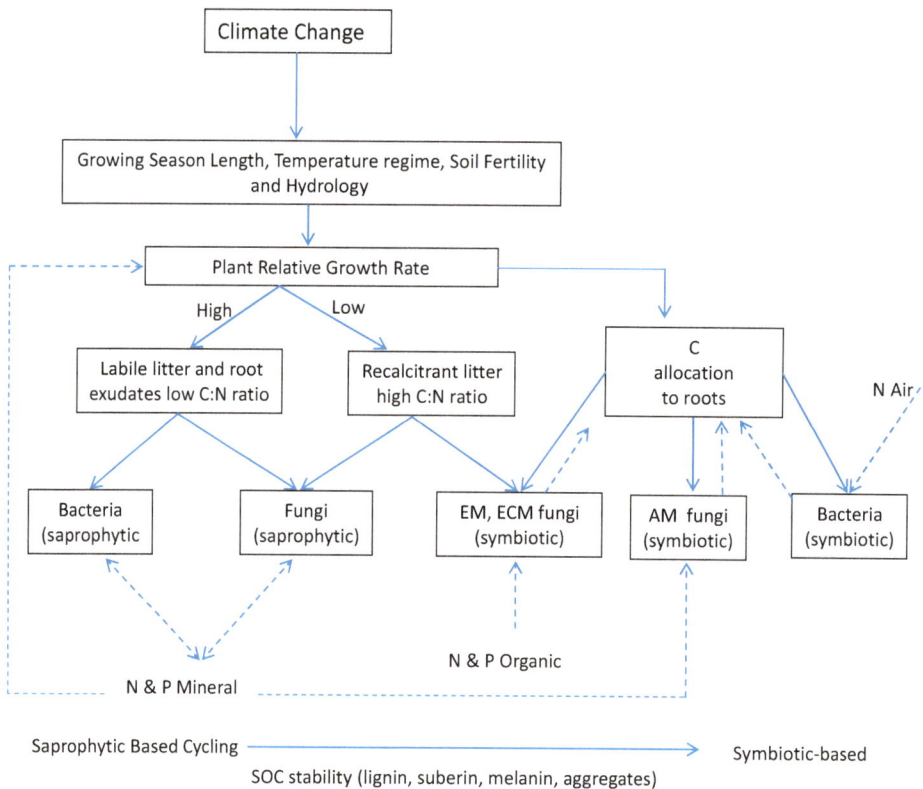

Fig. (5). The potential impact of climate change on soil microbes [9].

THE POTENTIAL IMPACT OF CLIMATE CHANGE ON SOIL MICROBES

Rainfall

A shift in the precipitation pattern due to climate change will change the soil profile's moisture content and oxygen level. The moisture and oxygen content levels decide the soil microbial biomass and their functions that alter the overall soil microbial community. Changes in rainfall patterns have the most extensive effects on community composition [15]. The small alteration in soil moisture content (30% reduction in water-holding capacity) can change species' composition in soil fungal communities. Nevertheless, the bacterial communities remain intact [16].

An Increased Concentration of Carbon Dioxide

The increase in the concentration of carbon dioxide levels not only increases methane efflux from the soil but also decreases the uptake of methane by up to 30% by soil microbes [17, 18]. Furthermore, the higher amount of carbon dioxide also results in an alteration in the microorganisms of plant leaves and leaves that decompose in streams. It could have great repercussions on the food chain because it acts as a source of nutrients for small phytophagous animals [19, 20]. The increased rate of respiration in soil microbial communities occurs due to the presence of elevated CO_2 in the atmosphere, which provides more carbon substrate for soil microorganisms [21].

Temperature

The elevation in temperature is probably to enhance the rate of fungal decomposition, resulting in an increased emission of carbon dioxide from the soil. However, an increase in temperatures also increases soil nitrogen levels, which has a reverse effect, *i.e.*, reduces the rates of fungal decomposition. The higher nitrogen availability in the soil negatively affects microbial activity and their composition in terms of diversity [19]. Nevertheless, the bacterial activities under stress exerted by an increased temperature occur less efficiently. Hence, these microorganisms absorb more carbon as carbon dioxide rather than transforming much of it into biomass [22]. The absorption of increased levels of carbon dioxide produced in this way and other ways stimulates plants to release N_2O and CH_4 [23]. Moreover, the overall response of microorganisms to warming for soil organic matter decay is dependent upon the temperature sensitivity of the decomposers and substrate availability. In general, global warming is probably due to an increase in the availability of nutrients in the soil due to the crop and to the increased rate of mineralization of soil organic matter by soil microbes [24].

Besides, soil warming due to increased temperatures increases the nitrogen mineralization rate, which will improve plant productivity. Apart from this, the plant community's diversification often changes with global warming [25 - 27]. The increased temperature and nutrient availability can lead to changes in the composition of flora [28, 29], which affect microorganisms' communities. At the same time, trees and shrubs that advance towards the north in the tundra under the influence of temperature alteration can also influence microbes in unknown ways through the shadows they cast on the ground [22].

CONCLUDING REMARKS

The soil microbial community is critical for the preservation of soil organic carbon and the supply of nutrients to plants. The importance of soil microbial communities in preserving healthy soil for future generations cannot be overstated. There is an urgent need to better understand the repercussions of climate change on key soil microorganisms and their impact on biogeochemical processes carried out in the soil and environment. We can use this information to make better preparedness for climate impacts and, ultimately, make a suitable design to combat further global warming and soil degradation.

CONSENT FOR PUBLICATION

Not applicable.

CONFLICT OF INTEREST

The author declares no conflict of interest, financial or otherwise.

ACKNOWLEDGEMENTS

Declared none.

REFERENCES

[1]   Lauber CL, Hamady M, Knight R, *et al.* Pyrosequencing-based assessment of soil pH as a predictor of soil bacterial community structure at the continental scale. Appl Environ Microbiol 2009; 75: 5111-20.
        [PMID: 19502440]

[2]   Talbot JM, Bruns TD, Taylor JW, *et al.* Endemism and functional convergence across the North American soil mycobiome. Proc Natl Acad Sci USA 2014; 111: 6341-6.
        [PMID: 24733885]

[3]   Goldmann K, Schöning I, Buscot F, *et al.* Forest management type influences diversity and community composition of soil fungi across temperate forest ecosystems. Front Microbiol 2015; 6: 1300.
        [PMID: 26635766]

[4]   Nacke H, Goldmann K, Schöning I, *et al.* Fine spatial scale variation of soil microbial communities under European beech and Norway spruce. Front Microbiol 2016; 7: 2067.

[http://dx.doi.org/10.3389/fmicb.2016.02067] [PMID: 28066384]

[5]   Tedersoo L, Bahram M, Põlme S, *et al.* Fungal biogeography. Global diversity and geography of soil fungi. Science 2014; 346: 1256688.
[http://dx.doi.org/10.1126/science.1256688] [PMID: 25430773]

[6]   Kaisermann A, de Vries FT, Griffiths RI, *et al.* Legacy effects of drought on plant-soil feedbacks and plant-plant interactions. New Phytol 2017; 215: 1413-24.
[PMID: 28621813]

[7]   Bahram M, Hildebrand F, Forslund SK, *et al.* Structure and function of the global topsoil microbiome. Nature 2018; 560: 233-7.
[PMID: 30069051]

[8]   Pugnaire FI, Morillo JA, Peñuelas J, *et al.* Climate change effects on plant-soil feedbacks and consequences for biodiversity and functioning of terrestrial ecosystems. Sci Adv 2019; 5: eaaz1834.
[http://dx.doi.org/10.1126/sciadv.aaz1834]

[9]   De Deyn GB, Cornelissen JH, Bardgett RD. Plant functional traits and soil carbon sequestration in contrasting biomes. Ecol Lett 2008; 11: 516-31.
[PMID: 18279352]

[10]   Bardgett RD, Wardle DA, Eds. Aboveground-belowground linkages: biotic interactions, ecosystem processes, and global change. Oxford: Oxford University Press 2010; p. 301.

[11]   Dijkstra FA, Hobbie SE, Reich PB. Soil processes affected by sixteen grassland species grown under different environmental conditions. Soil Sci Soc Am J 2006; 70: 770-7.

[12]   Glover J, Reganold JP, Bell LW, *et al.* Increased food and ecosystem security *via* perennial grains. Science 2010; 328: 1638-9.
[PMID: 20576874]

[13]   To JP, Zhu J, Benfey PN, *et al.* Optimizing root system architecture in biofuel crops for sustainable energy production and soil carbon sequestration. F1000 Biol Rep 2010; 2: 65.
[http://dx.doi.org/10.3410/B2-65] [PMID: 21173868]

[14]   Amend AS, Martiny AC, Allison SD, *et al.* Microbial response to simulated global change is phylogenetically conserved and linked with functional potential. ISME J 2016; 10: 109-18.
[PMID: 26046258]

[15]   Castro HF, Classen AT, Austin EE, *et al.* Soil microbial community responses to multiple experimental climate change drivers. Appl Environ Microbiol 2010; 76: 999-1007.
[PMID: 20023089]

[16]   Classen AT, Sundqvist MK, Henning JA, *et al.* Direct and indirect effects of climate change on soil microbial and soil microbial-plant interactions: What lies ahead? Ecosphere 2015; 6: 130.
[http://dx.doi.org/10.1890/ES15-00217.1]

[17]   Phillips RL, Whalen SC, Schlesinger WH. Influence of atmospheric CO_2 enrichment on methane consumption in a temperate forest soil. Glob Change Biol 2001; 7: 557-63.

[18]   Ineson P, Coward PA, Hartwig UA. Soil gas fluxes of N_2O, CH_4 and CO_2 beneath *Lolium perenne* under elevated CO_2: the Swiss free air carbon dioxide enrichment experiment. Plant Soil 1998; 198: 89-95.

[19]   Climate change could impact vital functions of microbes 2008. www.sciencedaily.com/releases/2008/06/080603085922.htm

[20]   Dutta H, Dutta A. The microbial aspect of climate change. Energy Ecol Environ 2016; 1: 209-32.

[21]   De Graaff MA, van Groenigen KJ, Six J, *et al.* Interactions between plant growth and soil nutrient cycling under elevated CO_2: a meta analysis. Glob Change Biol 2006; 12: 2077-91.

[22]   Zimmer C. The microbe factor and its role in our climate future 2010. https://e360.yale.edu/features/

the_microbe_factor_and_its_role_in_our_climate_future#:~:text=Within%20the%20planet's%20ocean
s%20and,the%20Earth's%20trees%20and%20plants.&text=Thanks%20to%20human%20activity%2C
%20carbon,planet%20can%20draw%20it%20down

[23]   Trinity College Dublin. Soil microbes accelerate global warming 2011. www.sciencedaily.com/
         releases/2011/07/110713131423.htm

[24]   Ruess L, Michelsen A, Schmidt IK, Jonasson S. Simulated climate change affecting microorganisms,
         nematode density and biodiversity in subarctic soils. Plant Soil 1999; 212: 63-73.

[25]   Harte J, Saleska S, Shih T. Shifts in plant dominance control carbon-cycle responses to experimental
         warming and widespread drought. Environ Res Lett 2006; 1: 014001.
         [http://dx.doi.org/10.1088/1748-9326/1/1/014001]

[26]   Walker MD, Wahren CH, Hollister RD, *et al.* Plant community responses to experimental warming
         across the tundra biome. Proc Natl Acad Sci USA 2006; 103: 1342-6.
         [PMID: 16428292]

[27]   Hoeppner SS, Dukes JS. Interactive responses of old-field plant growth and composition to warming
         and precipitation. Glob Change Biol 2012; 18: 1754-68.

[28]   Havström M, Callaghan TV, Jonasson S. Differential growth responses of *Cassiope tetragona*, an
         arctic dwarf-shrub, to environmental perturbations among three contrasting high and subarctic sites.
         Oikos 1993; 66: 389-402.

[29]   Hobbie SE. Temperature and plant species control over litter decomposition in Alaskan tundra. Ecol
         Monogr 1996; 66: 503-22.

CHAPTER 6

Prospects of Microbes in Organic Farming under the Scenario of Climate Change

Priyanka Chandra[1,*], **Parul Sundha**[1], **Rinki**[2], **Pooja Verma**[1], **Savitha Santosh**[3] and **Vanita Pandey**[2]

[1] *ICAR-Central Soil Salinity Research Institute, Karnal-132001, Haryana, India*

[2] *ICAR-Indian Institute of Wheat and Barley Research, Karnal-132001, Haryana, India*

[3] *ICAR-Central Institute for Cotton Research, Nagpur, Maharashtra-440010, India*

Abstract: Climate change is one of the minacious threats that is affecting agricultural production and food security the most. Agriculture is significantly involved in contributing to global warming with the use of chemical fertilizers. Soil microorganisms play an important role in several ecological processes in soil, such as the cycling of nutrients, nitrogen fixation, nitrification/denitrification, decomposition of organic matter, and mineralization/immobilization. These processes, carried out by microorganisms, are one of the most important components of organic farming. Climatic shifts are causing floods, droughts, and unseasonal rainfall and are showing potentially devastating effects on agricultural yields. Hence, there is an urgent need to develop strategies to make our farming systems more resilient to the consequences of climate change. This chapter presents the synergistic advantages of organic farming and the role of soil microbes, which could be effective climate change adaptation strategies for the agriculture sector, and will give information on the importance of soil microorganisms in organic farming.

Keywords: Climate change, Microbes, Mycorrhiza, Organic Farming.

INTRODUCTION

Climate change is one of the foremost threats to agricultural production and food security. Changes due to temperature, precipitation, and CO_2 concentration are the parameters that have a significant effect on crop growth and directly influence agricultural productivity [1]. Agriculture plays a crucial role in contributing to global warming with the use of chemical fertilizers. FAO has directly focused on organic farming as a novel alternative approach to channelize nutrients and energy

---

* **Corresponding author Priyanka Chandra:** ICAR-Central Soil Salinity Research Institute, Karnal-132001, Haryana, India; E-mail: priyanka.chandra921@gmail.com

Ashutosh Gupta, Shampi Jain & Neeraj Verma (Eds.)
All rights reserved-© 2022 Bentham Science Publishers

flow and to make the best use of renewable resources in the agro-world. A number of reports have revealed that emissions from organic farming are lower than those of conventional agriculture [2, 3].

These agricultural practises combat climate change by a) intensifying agro-ecological systems, b) diversifying cropping systems, c) varying livestock production, and d) educating farmers about the most effective eco-friendly practises [4]. Organic management practises can be very beneficial in reducing emissions of greenhouse gases (N_2O and NH_3) from cultivable lands. Several studies have shown that organic fertilization increases soil organic carbon in comparison to mineral fertilization, which means more sequestration of CO_2 from the atmosphere into the soil, which also provides additional benefits to biodiversity [5]. These parameters make organic farming more advantageous and possess a substantial potential for mitigating the consequences of climate change [6]. In organic farming systems, soil microorganisms play an important role in several ecological processes in soils, such as nitrogen fixation, nitrification/denitrification, decomposition of organic matter, and mineralization/immobilization, which are carried out by them and are one of the most important components of organic farming [7]. Climate change and extreme weather events such as floods, droughts, and unseasonal rainfall are occurring due to climate change and have major effects on agriculture. Hence, strategies need to be developed to make our farming systems more resilient to the repercussions of climate change. This chapter presents the synergistic advantages of microbes residing in soil and organic farming that could provide strategies for adaptation to the consequences of climate change in agriculture.

CLIMATE CHANGE *VS.* AGRICULTURE

Climate change is causing several extreme weather events that have a direct impact on agriculture production across the world. The rise in the mean seasonal temperature is leading to a change in the physiology of plants, which leads to a reduction in the time of crop maturity and, finally, yield reduction. It has been estimated that there will be a 3 to 16% decline in the productivity of crops by 2080 [8].

Since ancient times, Indian agriculture has been reliant on the seasonal monsoon, and changes in rainfall patterns are significantly affecting agricultural outputs. Indian agriculture is also severely affected by the increase in temperature. It has been reported that changes in pre-monsoon patterns lead to a loss in yield for the wheat crop, while severe droughts cause losses in rice production of an average of about 40% of total production. Wheat, soybean, mustard, groundnut, and potato yields are reduced by 3–7% when the temperature rises by 1 degree Celsius, and the loss increases as the temperature rises. The un-irrigated crops grown on 60% of the cropland in rain-fed areas are mostly affected by climate change. It has also been reported that in India, an increase of 0.5°C in the winter season is causing a reduction of 0.45 tonnes per hectare of wheat yield in rain-fed areas [9].

Due to climatic changes, agriculture in coastal areas has been damaged as fertile areas in these regions are highly sensitive to inundation and salinization [10]. Organic farming through microbes can be part of the solution to mitigate the effects of climate change because such practises reduce the use of chemical fertilizers and improve yields, along with reforming soil health and soil fertility and boosting carbon sequestration [11].

THE POTENTIAL BENEFITS OF ORGANIC FARMING

Organic farming practises demonstrate the potential for climate change resilience, agricultural production sustainability, and soil health [12]. Organic farming can sequester up to 32% of all greenhouse gas (GHG) emissions globally [13]. The potential benefits of organic farming are as follows:

- Organic farming increases carbon sequestration.
- It reduces greenhouse gas emissions.
- It reduces the financial risk as locally available renewable resources can replace the expensive chemical inputs.

Organic farming is a revitalization of a traditional system of agriculture in which there is no use of chemical fertilizers and pesticides while the use of manure, compost, and crop residues as a source of nutrients is being promoted. Several pesticides of biological origin are being used for pest management, and cropping system design is being emphasized [14]. Microorganisms have vital functions in organic farming as they play an important role in nutrient cycling and microbial communities. Bacterial and fungal communities dominate the soil ecosystem because of their role in decomposition and nutrient cycling, which are the indicators of soil health [15].

Mineral nutrients that are mineralized by microbes from organic matter are important. Thus, microbial biomass plays a dual role in mineralization and immobilization. The conversion of organic matter to inorganic nutrients (NH_4^+, NO_3^-, $H_2PO_4^-$, SO_4^{2-} and CO_2) is a process of mineralization, while immobilization is the process of immobilizing and making nutrients available for better absorption by plants for their maintenance and growth. Therefore, soils having high microbial communities have more potential for accumulation and scope of recycling of nutrients in the soil eco-system [14]. Microbial biomass in soils is higher in organic farming than in conventional farming because the supply of organic carbon content is greater in organic farming. Araújo and coworkers [16] also reported that the addition of organic products to the soil increases the soil microbial biomass.

Soil fertility is maintained in an organic farming system through increased microbial activity and high organic matter content. A high level of organic matter content slowly releases organic sources of nutrients, which are mobilized to plants by soil microbes, and the uptake of plant nutrients is enhanced. Organic farming, including legumes in crop rotation for biological nitrogen fixation, crop residues as a source of organic materials, livestock manures, and composts, has an impact on the farming system and the environment as well [17, 18]. Application of organic fertilizers and the addition of legumes in the crop rotation cycle influence the soil microbial population through better returns as crop residues. Crop-residues containing carbohydrates, cellulose, hemicelluloses, lignin, and proteins are excellent substrates for the growth of soil microbes. Carbohydrates mainly support the growth of soil bacteria, while others are degraded by fungi. However, all microbes are effective and play an important role in the degradation of crop residues [19].

The microbial biomass and their behaviour are also influenced by the carbon and nitrogen content in crop residues, wherein fresh crop residues have more carbon as compared to nitrogen [20, 21]. If C/N >20, then soil microbes accumulate carbon in their body walls and immobilize nitrogen, favouring an increase in microbial biomass. A sizable pool of active microbial strains forms the basis of the organic farming system and is required for the proper transformation of nutrients. Hence, the sufficient population of soil microbial communities and the maintenance of high organic matter content are important for a successful organic farming system [22].

ROLE OF MICROBES IN ORGANIC FARMING UNDER CHANGING CLIMATIC SCENARIO

Diverse populations of microorganisms and microfauna, along with an adequate amount of organic matter, are found in healthy soil, which is resistant to disturbances like tillage, flooding, drought, *etc.* Soil microbial populations play a very crucial role in the processes of an ecosystem, like the biochemical processes of C, N, S, and Fe. These elements act as essential nutrients for plants. Microbes carry out several chemical transformations and make unavailable forms of elements available to plants. The carbon cycle is one of the important biochemical cycles carried out by microbes [23]. The major and foremost function of the microbial community is the decomposition and transformation of organic substances. Fungi mainly secrete various degrading exo-and endo-enzymes that help in the decomposition and transformation of complex organic compounds. A healthy soil maintains all the key ecological functions, including the formation as well as decomposition of soil organic matter. Therefore, soil microbial activities are important for the improvement of soil quality [17, 24, 25].

Several bacteria and mycorrhizal fungi have been reported to stimulate plant growth and development. They can synthesize chemicals that limit pathogen damage to plants, for instance. *Trichoderma* is one type of fungal diversity that acts as a biocontrol agent in organic management. Many fungi, like *Penicillium*, *Aspergillus*, *Fusarium*, *etc.*, help crops by releasing phosphorus and other minerals from the soil. Several microbes like *Bacillus subtilis, B. megaterium, B. amyloliquefaciens, etc.*, are used in biological control and possess antagonistic activity. Mycorrhizae also act symbiotically and help plants convert the complex nutrients of the soil into available nutrients [26]. The mycorrhizal fungal hyphae expand into the soil to improve the porosity and structure. These hyphae absorb mineral nutrients and water from the soil and beyond the roots of plants and supply nutrients to them. Mycorrhiza demonstrates plant growth promotion properties through several mechanisms, which include the production of several plant growth promoting metabolites like amino acids, vitamins, and phytohormones. It has been reported by several researchers that mycorrhizal association also enhances plant tolerance to various biotic and abiotic stresses like drought, salinity, heavy metals, and provides protection from pathogens [27]. Plant growth-promoting bacteria, namely, *Pantoea* sp., *Chryseobacterium* sp., *Pseudomonas* sp., *Azospirillum brasilense, Bacillus subtilis, Rhizobium*, and *Azotobacter*, have the potential to act as potent bio-fertilizers as they have properties to solubilize nutrients like zinc and phosphorus and make them available to plants. PGPR can improve the extent or quality of plant growth directly and/or indirectly. Plant growth-promoting bacteria have been reported to enhance plant growth by several mechanisms and influence nutrient availability, growth, and yields. They are also found to influence soil conditions. Several plant growth-promoting bacteria belonging to these genera are *Azospirillum, Azotobacter, Pseudomonas, Klebsiella, Enterobacter, Alcaligenes, Arthrobacter, Burkholderia*, and *Serratia*. These bacteria colonize the host with high efficiency and produce a large number of plant growth metabolites [28] (Fig. **1**).

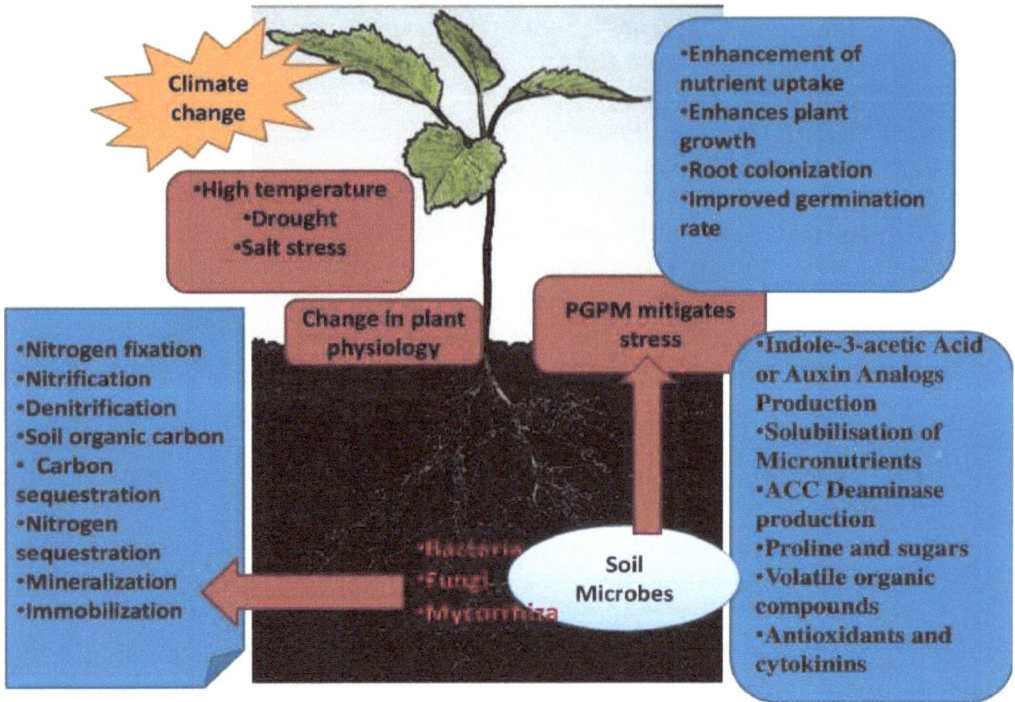

Fig. (1). Demonstration of the role of microbes in organic farming under the scenario of climate change.

EFFECT OF CLIMATE CHANGE ON SOIL MICROBIAL COMMUNITIES

Climate change is affecting soil microbial communities and their interactions with plants, directly or indirectly. These factors are directly altering the composition and functions of the ecosystem. DeAngelis and coworkers [29] reported that the bacterial to fungal ratio was enhanced in the soil microbial community with an increase in temperature by 5°C. This enhanced bacterial multiplication under raised temperatures, which is expected to bring changes in the soil ecosystem and the functions of soil organisms, is different in their functional properties.

As specific groups of soil microbes carry out nitrogen fixation, nitrification, and denitrification, and others carry out methanogenesis, so every group of microbes should be present in the ecosystem. If variations occur, they will directly or indirectly affect the ecosystem. However, in comparison to soil microbial communities, some processes, such as nitrogen mineralization, are more directly correlated with abiotic factors, which include soil moisture and temperature. Global warming affects the respiration rate of soil microbes as the microbial population and the processes they mediate are temperature sensitive, and it has received substantial attention from the scientific community across the world.

It has been reported that the diversity of arbuscular mycorrhizal fungi (AMF) influences the diversity of the plant population, also affects carbon and nitrogen sequestration simultaneously, and has an impact on soil properties, which include soil aggregation and moisture. Drought is also an extreme consequence of climate change, which is exacerbated as fungal and bacterial populations are temperature sensitive. It has been reported that changes in soil moisture availability (less than 30% reduction in water holding capacity of soil) lead to a shift of fungal communities in soil from one dominant group to another, while bacterial communities remain relatively constant [30]. This leads to the conclusion that bacteria are more adaptive to harsh conditions in comparison to fungi [31, 32]. Studies revealed that under drought conditions, AMF loses its potential to form a symbiotic relationship with plants. However, under similar conditions, plant growth-promoting bacteria such as *Pseudomonas* sp., Actinobacteria, and Deltaproteobacteria protect plants and persuade different processes, which include the increased cellular division in roots and root hairs, so that the number of root hairs is increased. The change in the physiology of plants leads to more water absorption and enhanced nutrient uptake by plant roots, especially from deeper layers of soil. Under drought conditions, plant growth-promoting bacteria may further regulate the levels of stress-related hormones, *i.e.* abscisic acid and ethylene, potentially leading to better plant drought tolerance.

Mycorrhizal fungi form an association with almost all terrestrial vegetation, and approximately 90% of plants form the mycorrhizal association, making them an important stakeholder in biochemical cycling and processes. Mycorrhizal fungi exchange nutrients for plant carbon (mostly phosphate), thus influencing plant carbon to nutrient ratios and, subsequently, plant productivity. The mycorrhizal composition can change with climatic factors like temperature [33]. Similarly, Hawkes and coworkers [34] postulated that an increase in temperature increases carbon allocation to mycorrhizal hyphae. This turns a symbiotic mycorrhizal association into a parasitic one. Some favourable results of climate change have also been observed. Several studies have found that increased CO_2 levels improve the interaction of legumes and rhizobia [35, 36]. Montealegre and coworkers [37] also revealed that with atmospheric CO_2 enrichment, *Rhizobium leguminosarum* growth and interaction with legumes increased in comparison to other strains.

CONCLUDING REMARKS

Microbes carry out a number of critical functions in the ecological system. Organic farming includes several agricultural practices, which include the use of farmyard manure and compost. The retention of crop residues, mulching, and legume cultivation and such components have a positive impact on soil health, which includes improved soil organic content and microbial biomass. The prediction of microbial diversity is quite a complex process, but a rich microbial diversity is generally perceived under organic farming. The effects of climate change on microbial populations and carbon dynamics remain unclear, but their interactions are affected by climatic fluctuations. The effects of climate change on soil carbon dynamics are also complex. However, these factors are needed to focus on under a climate change scenario. Climate change also induces adaptation processes in plants and microbial communities. Microbes, which possess plant growth-promoting properties such as mycorrhizae and nitrogen-fixing bacteria, support plant growth through several mechanisms and also maintain soil health and quality, which are also important in organic farming under climate change. It can be concluded that facets of organic farming could be a better option for sustainable agriculture in a changing climatic scenario.

CONSENT FOR PUBLICATION

Not applicable.

CONFLICT OF INTEREST

The author declares no conflict of interest, financial or otherwise.

ACKNOWLEDGEMENTS

Declared none.

REFERENCES

[1]   Gadgil S. Climate change and agriculture – an Indian perspective. Curr Sci 1995; 69: 649-59.

[2]   Mondelaers K, Aertsens J, van Huylenbroeck G. A meta-analysis of the differences in environmental impacts between organic and conventional farming. Br Food J 2009; 111: 1098-119.

[3]   Aguilera E, Guzmán G, Alonso A. Greenhouse gas emissions from conventional and organic cropping systems in Spain. I. Herbaceous crops. Agron Sustain Dev 2015; 35: 713-24.

[4]   Two essays on climate change and agriculture: a developing country perspective. FAO Social and Economic Development, Food and Agriculture Organization, Rome, Paper No 145 2000. www.fao.org/3/x8044e/x8044e00.htm

[5]   Barakat MR, Yehia TA, Sayed BM. Response of newhall naval orange to bio-organic fertilization under newly reclaimed area conditions I: Vegetative growth and nutritional status. J Hortic Sci Ornam Plants 2012; 4: 18-25.

[6]   FAO. Organic agriculture and climate change mitigation. A report of the round table on organic agriculture and climate change. December 2011, Rome, Italy 2011. www.fsnnetwork.org/sites/default/files/organic_ agriculture_and_climate_change_miti.pdf

[7]   Araújo ASF, Melo WJ. Soil microbial biomass in organic farming system. Cienc Rural 2010; 40: 2419-26.

[8]   Parry ML, Canziani OF, Palutikof JP, *et al.* Climate Change 2007: Impacts, Adaptation and Vulnerability. Cambridge University Press, Cambridge, UK 2007; p. 976.

[9]   Khan SA, Sanjeev K, Hussain MZ, *et al.* Climate change, climate variability and Indian agriculture: impacts vulnerability and adaptation strategies.Climate change and crops, environmental science and engineering. Singh SN, Ed. Berlin, Heidelberg: Springer-Verlag 2009; pp. 19-38.

[10]   Mahato A. Climate change and its impact on agriculture. Internat J Sci Res Pub 2014; 4: 1-6.

[11]   Goh KM. Greater mitigation of climate change by organic than conventional agriculture: A Review. Biol Agric Hortic 2011; 27: 205-29.

[12]   Meng L, Ding WCZ. Long-term application of organic manure and mineral fertilizers on aggregation and aggregate-associated carbon in a sandy loam soil. Soil Tillage Res 2012; 124: 170-7.

[13]   Kotschi J, Müller-Sämann K. The role of organic agriculture in mitigating climate change - a scoping study. Bonn: IFOAM 2004; p. 64.

[14]   Lori M, Symnaczik S, Mäder P, *et al.* Organic farming enhances soil microbial abundance and activity-A meta-analysis and meta-regression. PLoS One 2017; 12: e0180442.
[http://dx.doi.org/10.1371/journal.pone.0180442] [PMID: 28700609]

[15]   Zarb J, Ghorbani R, Koocheki A, *et al.* The importance of microorganisms in organic agriculture. Outlooks Pest Manag 2005; 16: 52-5.

[16]   Araújo ASF, Santos VB, Monteiro RTR. Responses of soil microbial biomass and activity for practices of organic and conventional farming systems in Piauí state. Brazil Eur J Soil Biol 2008; 44: 225-30.

[17]   Mäder P, Fliessbach A, Dubois D, *et al.* Soil fertility and biodiversity in organic farming. Science 2002; 296: 1694-7.
[PMID: 12040197]

[18]   Tinker PB. Organic farming - nutrient management and productivity. United Kingdom: The Intenational Fertiliser Society, Proceedings. York 2001; 471: pp. 1-24.

[19]   Rahman MM, Islam AM, Azirun SM, *et al.* Tropical legume crop rotation and nitrogen fertilizer effects on agronomic and nitrogen efficiency of rice. ScientificWorldJournal 2014; 2014: 490841.
[http://dx.doi.org/10.1155/2014/490841] [PMID: 24971378]

[20]   Wang J, Sainju UM. Soil carbon and nitrogen fractions and crop yields affected by residue placement and crop types. PLoS One 2014; 9: e105039.
[http://dx.doi.org/10.1371/journal.pone.0105039] [PMID: 25119381]

[21]   Du Z, Xie Y, Hu L, *et al.* Effects of fertilization and clipping on carbon, nitrogen storage, and soil microbial activity in a natural grassland in southern China. PLoS One 2014; 9: e99385.
[http://dx.doi.org/10.1371/journal.pone.0099385] [PMID: 24914540]

[22]   Araújo ASF, Leite LFC, Santos VB, *et al.* Soil microbial activity in conventional and organic agricultural systems. Sustainability 2009; 1: 268-76.

[23]   Hartmann M, Frey B, Mayer J, *et al.* Distinct soil microbial diversity under long-term organic and conventional farming. ISME J 2015; 9: 1177-94.
[PMID: 25350160]

[24]   Esperschütz J, Gattinger A, Mäder P, *et al.* Response of soil microbial biomass and community structures to conventional and organic farming systems under identical crop rotations. FEMS Microbiol Ecol 2007; 61: 26-37.
[PMID: 17442013]

[25]   Stark CH, Condron LM, O'Callaghan M, *et al.* Differences in soil enzyme activities, microbial community structure and short-term nitrogen mineralisation resulting from farm management history and organic matter amendments. Soil Biol Biochem 2008; 40: 1352-63.

[26]   Augé RM, Toler HD, Saxton AM. Arbuscular mycorrhizal symbiosis alters stomatal conductance of host plants more under drought than under amply watered conditions: a meta-analysis. Mycorrhiza

2015; 25: 13-24.
[PMID: 24831020]

[27]   Augé RM, Toler HD, Saxton AM. Arbuscular mycorrhizal symbiosis and osmotic adjustment in response to NaCl stress: a meta-analysis. Front Plant Sci 2014; 5: 562.
[http://dx.doi.org/10.3389/fpls.2014.00562] [PMID: 25368626]

[28]   Yasin NA, Khan WU, Ahmad SR, *et al.* Imperative roles of halotolerant plant growth-promoting rhizobacteria and kinetin in improving salt tolerance and growth of black gram (*Phaseolus mungo*). Environ Sci Pollut Res Int 2018; 25: 4491-505.
[PMID: 29185225]

[29]   DeAngelis KM, Pold G, Topçuoğlu BD, *et al.* Long-term forest soil warming alters microbial communities in temperate forest soils. Front Microbiol 2015; 6: 104.
[http://dx.doi.org/10.3389/fmicb.2015.00104] [PMID: 25762989]

[30]   Malik AA, Chowdhury S, Schlager V, *et al.* Soil fungal: bacterial ratios are linked to altered carbon cycling. Front Microbiol 2016; 7: 1247.
[http://dx.doi.org/10.3389/fmicb.2016.01247] [PMID: 27555839]

[31]   Ushio M, Miki T, Balser TC. A coexisting fungal-bacterial community stabilizes soil decomposition activity in a microcosm experiment. PLoS One 2013; 8: e80320.
[http://dx.doi.org/10.1371/journal.pone.0080320] [PMID: 24260368]

[32]   Davies LO, Schäfer H, Marshall S, *et al.* Light structures phototroph, bacterial and fungal communities at the soil surface. PLoS One 2013; 8: e69048.
[http://dx.doi.org/10.1371/journal.pone.0069048] [PMID: 23894406]

[33]   Rillig MC, Mummey DL. Mycorrhizas and soil structure. New Phytol 2006; 171: 41-53.
[PMID: 16771981]

[34]   Hawkes CV, Hartley IP, Ineson P, *et al.* Soil temperature affects carbon allocation within arbuscularmycorrhizal networks and carbon transport from plant to fungus. Glob Change Biol 2008; 14: 1-10.

[35]   Compant S, van der Heijden MG, Sessitsch A, *et al.* Climate change effects on beneficial plant-microorganism interactions. FEMS Microbiol Ecol 2010; 73: 197-214.
[PMID: 20528987]

[36]   Wang ZG, Bi YL, Jiang B, *et al.* Arbuscular mycorrhizal fungi enhance soil carbon sequestration in the coalfields, northwest China. Sci Rep 2016; 6: 34336.
[http://dx.doi.org/10.1038/srep34336] [PMID: 27748365]

[37]   Montealegre C, Van Kessel C, Russelle M, *et al.* Changes in microbial activity and composition in a pasture ecosystem exposed to elevated atmospheric carbon dioxide. Plant Soil 2002; 243: 197-207.

CHAPTER 7

Dynamic Interplay of Soil and Microbes for Sustainable Ecological Balance

Ashwini A. Waoo[1,*] and **Shivangi Agnihotri**[1]

[1] Department of Biotechnology, Faculty of Life Sciences and Technology, AKS University, Satna, MP, India

Abstract: A microorganism is a term given to small living beings whose size is measured in microns. Bacteria, fungi, algae, and protozoans are a few of them that reside in the air, water, and soil. This review is about the microorganisms found in soil. These microorganisms have different functions in soil decomposition of dead organic matter, such as ecological food web balance, and making nutrients available to plants. Recently, their role in alleviating different abiotic stresses like salinity and drought has been marvelous. These microbes are also being used in biopesticide form, which is environmentally friendly and safe for other living organisms. Bacteria convert the inaccessible nutrients from dead matter into usable forms. Actinomycetes give off the typical smell of soil, and these microorganisms are also being used as a source of therapeutic medicines. Fungi are helpful in the way that they break down impossible nutrients, which are then available to other microbes. They also colonize plant roots and thus aid in plant growth. Algae promote submerged aeration as their photosynthesis is faster and adds more oxygen. Algae also help prevent the loss of nitrates that help in building soil structures by promoting the weathering of rocks. Nematodes help maintain the ecological equilibrium of their habitat. Viruses are the mode of gene transfer between organisms in the soil. Thus, these microorganisms have different functions in the soil to maintain the soil's structure and the balance between the environment and its living beings.

Keywords: Algae, Bacteria, Fungi, Microorganisms, PGPR, Soil.

INTRODUCTION

Soil microorganisms are the micro-sized organisms found in soil. These organisms define the properties of a particular soil with their different functions. Bacteria, protozoa, actinomycetes, algae, fungi, and viruses are the various types of microorganisms present in the soil. The groups are categorized based on their unique characteristics and their functions in the soil. These organisms always exist in coordination with each other.

* **Corresponding author Ashwini A. Waoo:** Department of Biotechnology, Faculty of Life Sciences and Technology, AKS University, Satna, MP, India; E-mail: ashwiniwaoo@gmail.com

Ashutosh Gupta, Shampi Jain & Neeraj Verma (Eds.)

There are various types of microorganisms found in soil. They are discussed below in brief.

TYPES OF MICROORGANISMS FOUND IN SOIL

Bacteria

Bacteria are microscopic organisms that constitute a large domain of prokaryotic microorganisms. Bacteria have only one cell. These bacteria are present in a very large number, *i.e.*, around one billion, in a small portion of moist, fertile soil. Bacteria can be decomposers that depend on dead and decaying plant material as well as any organic waste, releasing nutrients. Bacteria convert the inaccessible nutrients from dead matter into usable forms.

Actinomycetes

Actinomycetes are soil microorganisms, also called "thread or ray bacteria". These organisms have characteristics similar to both bacteria and fungi, and thus they are called "connecting links" between these two organisms. The typical smell of a particular soil is due to actinomycetes. These microorganisms are also being used as a source of therapeutic medicines.

Fungi

These are unique organisms, as they do not resemble plants or animals. The basic structure of fungi includes fibrous strings called hyphae. These hyphae, collectively called mycelium, have less than 0.8 mm in width and approximately one meter in length. Some of the fungi are useful, and some are harmful to other soil microbes. Fungi are helpful in breaking down impossible nutrients, which are then available to other microbes. Some of the fungi also colonize plant roots and thus help in plant growth. This type of relationship, which is equally beneficial to both, is called mycorrhiza. The plant benefits by receiving needed nutrients, and the fungi benefit from carbohydrates from the plant. Fungi can also show parasitic behaviour and obtain their food by getting attached to plants or other organisms for survival reasons.

Algae

It is a group of oxygenic, photosynthetic, and eukaryotic microorganisms. These algae are found mainly in moist soils that have a lot of sunlight. In one gram of soil, 100 to 10,000 algal organisms can be measured. Algae are photosynthetic, *i.e.*, they synthesize their food by using carbon dioxide from the atmosphere and sunlight for energy.

Role of Algae in the Soil

- Algae increase soil fertility by the addition of humus.
- The decomposition of dead algae increases the carbon content of the soil.
- Algae reduce soil erosion by increasing the water-holding capacity of the soil.
- Algae promote submerged aeration as their photosynthesis process is faster and adds more oxygen.
- Algae help in preventing the loss of nitrates and in building soil structures by promoting the withering of rocks.

Protozoans

These are colorless, single-celled animal-like organisms. Their size varies from a few microns to a few millimeters and is larger than bacteria. Their numbers in arable soil range from 10,000 to 100,000/gram of soil, and they are plentiful in the surface layer of soil. They have a dormant stage in their life cycle, which helps them withstand adverse soil conditions. Their salient features are that they help in maintaining microbial equilibrium by feeding soil bacteria, acting as biocontrol agents, and also causing disease in humans, *e.g.*, amoebic dysentery.

Viruses

Viruses are infectious agents that cannot thrive or reproduce outside of a host body. Viruses are incredible as they are capable of transferring genes from one host to another and also maintaining the mortality rate of soil organisms. Thus, they maintain the soil ecology. Soils are found to be the home of various viral species.

Nematodes

Nematodes are often referred to as roundworms. They are multicellular with an unsegmented body. They range from 50 microns in width to a one-millimeter length. Nematodes help maintain the ecological equilibrium of their habitat as they are found to function at a different level in the food web. At their first level, they feed on the plants and algae. Those in the second level feed on bacteria and fungi, and at a higher level, some feed on other nematodes.

Free-living nematodes are classified into groups as bacterial feeders, which consume bacteria, and fungal feeders, which feed on fungus by puncturing their cell walls and sucking out the internal content. Nematodes that feed on others in their community (*i.e.*, nematodes) and also on protozoa are called predatory nematodes. These nematodes are attached to the cuticles of larger nematodes and take extracts from their internal body parts. When nematodes consume bacteria and fungi, ammonium is released, and thus these nematodes help in releasing nutrients. Nematodes are also indicators of soil [1].

SIGNIFICANT ROLE OF SOIL MICROBES

Microorganisms play an essential role in the soil to sustain plant life and ecological balance (Fig. 1) [2]. Microorganisms play an important role in the decomposition of organic matter. All the microorganisms present in the soil break down organic matter. This breakdown of organic matter is the result of several chemical reactions, which include chemical modification of matter, physical breakdown, and the final release of nutrients.

Decomposition of Organic Matter

This fragmentation of organic matter is considered a biological process as it includes chemical and physical changes by soil organisms such as microbes, microarthropods, earthworms, ants, and beetles. These organisms work at different levels of the decomposition process. As soon as an organism or part of it dies, the decomposition process starts. Microorganisms present in organic matter use enzymes for oxidization and obtain energy as well as carbon. Earthworms and other larger soil animals, such as mites and ants, break organic matter into smaller particles. Thus, the surface area available for microorganisms to colonize and decompose the organic matter has also increased.

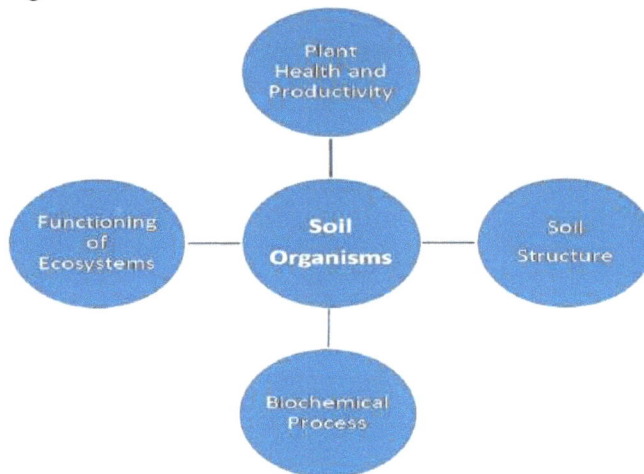

Fig. (1). Soil microorganism's functions.

The organic material is further decomposed to release nutrients into the soil. Different compounds break at different times according to their ease. Firstly, the breakdown of sugars and amino acids takes place, followed by cellulose and lignin. The phenols and waxes remain in the soil for the longest duration.

Recycling of Nutrients

Soil microbes play a crucial role in returning nutrients to their mineral forms, which plants can take up again. This process is known as mineralization.

These soil microbes behave as decomposers and break down dead plant and animal matter. Imagine a world without decomposers. We can see everywhere the huge collection of dead matter. If these recyclers were absent, the world would become a dump yard of dead matter. Thus, these microbes do a very important job of breaking down dead matter into simple chemical compounds and enabling plants to access the nutrients they require for their survival. The microbes involved in decomposition use organic carbon in the dead matter as their source of energy. Other nutrients released in this process, such as nitrogen, potassium, and phosphorus, are utilized by plants for their survival. These microbes do a very important job of decomposition and recycling of nutrients, thus converting the world's garbage into the building blocks of living things [3].

Formation of Humus

The dark brown, jelly-like substance obtained after the complete breakdown of organic matter by the action of microbes is called humus. This humus has remained unchanged for over a millennium. Soil retains moisture with the help of humus content, which suppresses plant diseases. Humus molecules have negatively charged ions on their surface, which bind to the plant nutrients carrying positively charged ions, thus enabling the soil's cation exchange capacity.

Creating Soil Structure

The polysaccharides, gums, and glycoproteins are secreted by microbes that join soil minerals together, forming the basis for soil structure. Further soil aggregates are formed and bound well by fungal hyphae and plant roots. Soil structure is essential to good plant growth. Organic matter and microbes help in the arrangement and holding of aggregates of soil [4]. Good soil is defined by better soil porosity and aeration at the optimum level for plants and microbes residing in the soil. Thus, the soil aggregates that are required are stable for wetting. Bacteria and fungi secrete a variety of mucilaginous polysaccharides that help to form clay, sand, organic matter, and soil aggregates. These polysaccharides are produced by utilizing carbon from fresh plant material. Fungal hyphal networks encourage enmeshment of soil particles with organic residue and thus form macroaggregates (250–2,000 mm) [5]. Decomposing plant residues and microbial debris encrusted with soil particles by mucilage form the core of microaggregates (20–250 mm) [6].

As the decomposition of organic material progresses, microbial metabolites further permeate the surrounding mineral crust, increasing the interparticle cohesion and promoting the stability of aggregates. The quality of soil is considered best when it provides a suitable living condition for different microbes. The soil should provide a mode for the exchange of ionic reactions and substrate availability.

Nitrogen Fixation

The process of the conversion of atmospheric molecular nitrogen (N_2 gas) in the air to ammonia or various nitrogenous compounds in the soil is called nitrogen-fixation. The bacteria involved in this process are called nitrogen-fixing bacteria. This nitrogen fixation is a crucial step in agriculture, as with this the availability of nitrogen to crops is increased to a greater extent.

Nitrogen-fixing bacteria are categorized into two types. The first category is of non-symbiotic bacteria and includes the cyanobacteria (*Anabaena* and *Nostoc*) and genera such as *Beijerinckia*, *Azotobacter*, and *Clostridium*. The second category is of those bacteria that live in attached forms with plants and show a mutual relationship with them. They are termed symbiotic bacteria. An example of symbiotic bacteria is *Rhizobium*, which is accompanied by leguminous plants of the pea family. Another example is *Frankia*, which is found to be attached to some dicotyledonous species and *Azospirillum* species, which are found to be attached to certain cereal grasses.

The host plant's root hairs are colonized by nitrogen-fixing symbiotic bacteria, and here at root nodules, these bacteria multiply and initiate the nodule formation by enlarging the plant cells in intimate association. The nodules act as a reaction chamber where the free nitrogen gets converted to ammonia, which is used up by the plants for their growth. Soils lacking these bacteria are inoculated with soaked seeds in the symbiotic bacterial culture that is sown for nodule formation (*e.g.* beans, clovers, peas, alfalfa, soybeans, *etc.*).

Plant Growth Promotion

Some soil microorganisms produce various substances that promote plant growth, including different hormones such as auxins, gibberellins, and antibiotics. Rhizosphere microorganisms have been shown to enhance the growth of many different crops in saline soil and some in drought-affected areas. This approach is helpful as the development of salt-tolerant germplasm is tedious work and costly, and it is also not possible everywhere. The bacteria that help in plant growth are termed plant growth-promoting rhizobacteria (PGPR). PGPR was first introduced and defined by Kloepper and Schroth [7]. They described PGPR as soil bacteria having the ability to colonize the roots of plants after inoculation into seeds, and they enhance plant growth. These microorganisms promote plant growth by an increase in nutrient uptake and the prevention of various diseases by destroying heavy metal ions and degrading xenobiotic compounds [8].

These microorganisms help to alleviate biotic stress in affected crops. This use of microorganisms is a safe and ecological alternative to sustainable agriculture. The microbes acting as PGPR are perfect research tools to investigate the mechanisms of tolerance to different stresses. They can also decode adaptation and response mechanisms of different crops [9].

Control of Pests and Diseases

Chemical pesticides are widely used in modern agriculture to control pests and increase crop yields. In developing countries, their use has skyrocketed in recent decades. These pesticides are made of synthetic chemicals that, when released into the soil, cause serious side effects on fauna and flora and also kill helpful microorganisms and predators. They are also harmful to humans and domestic animals. Their residues contaminate groundwater. In contrast to them, biopesticides are biological chemicals that control pests through a safe mechanism. These biopesticides include living organisms and their secretory compounds such as phytochemicals or physio-chemicals that are equally toxic to the pest-harming crops. These biopesticides are ecologically safe and are harmless to humankind and the surrounding environment. Some examples are *Trichoderma* (bioherbicides) for *Phytophthora* and bioinsecticides, *e.g.*, *Bacillus thuringiensis* and *B. sphaericus*.

The commercial production of the soil bacterium *Bacillus thuringiensis* (Bt) for controlling caterpillars, which cause harm to crops. Beetles and flies are also controlled by some strains of Bt. Several strains of *Trichoderma* are also commercially available to control root diseases.

CONCLUDING REMARKS

Microorganisms in soil are small-sized organisms that play a major role in the structuring of soil by decomposing dead and organic matters, making nutrients accessible to plants, promoting plant growth and controlling harmful pests of crops. The decomposition reaction performed by microorganisms helps to clean up the environment; otherwise, this dead and decaying matter would pile up and become a huge problem to live with. This fragmentation process also maintains the nutrient cycle in nature. Thus, microorganisms are an important component in maintaining ecological equilibrium.

The use of microorganisms as biopesticides is the most important finding, as in recent times, with the increasing demand for food and crops, the use of harmful and toxic chemical fertilizers is increasing to a large extent. The use of biopesticides is the safest alternative to chemical fertilizers. Beacuse these pesticides are harmless to mankind and the surrounding environment.

There are various and innumerable varieties of microorganisms that are still unknown. Thus, research should be extended to find more useful microorganisms for mankind and the environment.

CONSENT FOR PUBLICATION

Not applicable.

CONFLICT OF INTEREST

The author declares no conflict of interest, financial or otherwise.

ACKNOWLEDGEMENTS

Declared none.

REFERENCES

[1]   Johns C. Living Soils: The Role of Microorganisms in Soil Health. Australia: Future Directions International Pty Ltd. 2017; p. 7.

[2]   Furtak K, Gajda AM. Activity and variety of soil microorganisms depending on the diversity of the soil tillage system. In: De Oliveira A, Ed. Sustainability of Agroecosystem. Intech Open, UK 2017; p. 18.
        [http://dx.doi.org/10.5772/intechopen.72966]

[3]   Jacoby R, Peukert M, Succurro A, *et al.* The role of soil microorganisms in plant mineral nutrition-current knowledge and future directions. Front Plant Sci 2017; 8: 1617.
        [http://dx.doi.org/10.3389/fpls.2017.01617] [PMID: 28974956]

[4]   Tisdall JM, Oades JM. Organic matter and water-stable aggregates in soils. Eur J Soil Sci 2006; 33: 141-63.

[5]   Gupta VVSR, Germida JJ. Distribution of microbial biomass and its activity in different soil aggregate size classes as affected by cultivation. Soil Biol Biochem 1988; 20: 777-87.

[6]   Six J, Bossuyt H, Degryze S, *et al.* A history of research on the link between (micro) aggregates, soil biota, and soil organic matter dynamics. Soil Tillage Res 2004; 79: 7-31.

[7]   Kloepper JW, Schroth MN. Plant growth-promoting rhizobacteria on radishes. Proc 4th Int Conf Plant Path Bact Angers 1978; pp. 879-82.

[8]   Hayat R, Ali S, Amara U, *et al.* Soil beneficial bacteria and their role in plant growth promotion: a review. Ann Microbiol 2010; 60: 579-98.

[9]   Grover M, Ali SkZ, Sandhya V, *et al.* Role of microorganisms in adaptation of agriculture crops to abiotic stresses. World J Microbiol Biotechnol 2011; 27: 1231-40.

CHAPTER 8

Degradation of Hazardous Organic and Inorganic Waste by Microorganisms

Arbind Kumar Gupta[1,*], Deo Kumar[2] and Atik Ahamad[3]

[1] *Department of Soil Science, College of Forestry, Banda University of Agriculture & Technology, Banda-210001 U.P., India*

[2] *Department of Soil Science & Agril Chemistry, College of Agriculture, Banda University of Agriculture & Technology, Banda-210001, U.P., India*

[3] *KVK, Bharari, Jhansi, U.P., India*

Abstract: With the increasing population of the world and the daily life demands supplied through industries and modern industrialized agricultural systems, the need for the preservation of ecosystems is increasing day by day. Many industrial processes result in large amounts of organic waste as well as inorganic contaminants that deteriorate food and water quality. Immediate measures to avoid the negative impact on the environment are necessary. The generation of large quantities of hazardous materials in the form of heavy metals, radioactive substances, phenolic compounds, and volatile organic chemicals has resulted in the requirement for new and environmentally safe methods for their elimination. *In situ* degradation of hazardous organic materials by microbes is often the most cost-effective clean-up approach. Biological treatment of these hazardous wastes is potentially effective, practical, and economical. Bioremediation is measured as one of the safer, cleaner, cost-effective, and eco-friendly technologies for decontaminating sites. It uses numerous agents such as bacteria, yeast, fungi, algae, and higher plants as its main tools in treating oil spills, pesticides, radionuclides, polluted groundwater, and heavy metals existing in the environment. Currently, different methods and strategies are being applied in different parts of the world. Phytoextraction, biostimulation, fungal bioremediation, and rhizofiltration are some of the more common ones. Because of specific applications, all bioremediation techniques have their advantages and disadvantages.

Keywords: Biodegradation, Biological pesticides, Bioremediation, Pollution.

INTRODUCTION

The rapid increase in the world population in the last decades, accompanied by the intensification of human activities, has brought serious environmental problems such as pollution of soil, water, air, *etc.* In the future, negative impacts

*** Corresponding author Arbind Kumar Gupta:** Department of Soil Science, College of Forestry, Banda University of Agriculture & Technology, Banda-210001 U.P., India; E-mail: arbind4gupta@gmail.com

Ashutosh Gupta, Shampi Jain & Neeraj Verma (Eds.)
All rights reserved-© 2022 Bentham Science Publishers

may cause severe global climatic problems and might be a menace to the existence of the human race. Soil is a natural and living thing that contains several microorganisms, *i.e.*, microflora and microfauna. It gives mechanical as well as nutritional support to plants growing in the environment.

Several pollutants are present in our environment that pollute soil, air, water, *etc.* Soil contamination by pollutants like heavy metals is a serious environmental issue and poses significant impacts on the ecosystem. Some heavy metals are essential trace elements, but most of them can be toxic to all forms of life at high concentrations due to the formation of complex compounds within the cell.

Soil contaminants like heavy metals are widely used in different industries like textile, leather, paper, electroplating, chrome plating, petroleum refining, paint, and fabricating industries. The waste discharge from these industries contains a huge quantity of toxic substances, and untreated effluents cause serious ecological and environmental contamination [1]. Deposition and persistence of these organic pollutants, *i.e.*, sewage and sludge, in the ecosystem cause environmental contamination. Heavy metals discharged from industries into the water bodies undergo several chemical transformations, creating a huge impact on the environment and human health. Heavy metals present in the soil system subsequently enter into the food chain and ultimately cause a heavy risk to human health. Every year, a massive amount of contaminants such as urban and industrial waste are dumped into the environment. Approximately, 6000 metric tonnes of chemical compounds have been synthesized up to this point, with 1,000 new chemicals synthesized each year [2]. Moreover, injudicious use of agrochemicals like phosphate fertilizers, which show a high load of heavy metals like cadmium (Cd) and fluorine (F), insecticides, and pesticides, the amounts of heavy metals deposited onto the surface of the Earth are, however, many times greater than depositions from natural sources. The dumping of such hazardous waste into the environment is extremely harmful to living entities. The contaminants in the environment cause ecological imbalances and are a major global concern today. At present, a number of new technologies have been developed worldwide that emphasize the destruction of environmental pollutants rather than the conventional approach of disposal.

In agricultural practices for crop production, many harmful inorganic chemicals, fertilizers, are used to control insects, pests, diseases, and weeds. These are applied to plants as well as to soil. These chemicals are, however, normally detoxified in the soil within a reasonable time if used in the recommended amount. The detoxification of such poisonous chemicals is done by soil microorganisms. Soil organisms decompose these unusual chemicals and save the plants from their adverse effects. The transformation of heavy metals and pesticides by microbes reduces their toxic levels through different mechanisms.

Bioremediation involves enhanced degradation of various toxic compounds by transforming them into non-toxic substances, specifically carbon dioxide and water. This process can be carried out either *in situ* by taking advantage of indigenous microbes or by the introduction of bacterial or fungal strains to achieve complete detoxification of hazardous compounds. This chapter discusses various methods such as fungal bioremediation, phytostabilization, rhizofiltration, phytoextraction, phytovolatilization, and phytotransformation in bioremediation practices.

SOURCES OF POLLUTION

Pollution by Pesticides

Pesticides are the chemicals being used to control harmful organisms since they have become an essential tool in modern agriculture. Pesticides include insecticides, fungicides, herbicides, rodenticides, and nematicides. All chemicals are harmful when their concentrations are high enough, even though they are beneficial at low concentrations. Chlorinated pesticides are highly persistent, and the typical value is 3 to 4 years, *e.g.*, DDT: 4 years, BHC: 3 years, *etc.* The persistence of organophosphate and carbamate insecticides in the soil is low. Pesticides have an adverse impact on soil flora and soil fauna, affecting the populations of arthropods (insects, mites, spiders, millipedes, and centipedes), molluscs (snails and slugs), earthworms, and protozoa, which are very important in maintaining soil structure and plant nutrition. Ideal pesticides should not pollute the environment because they should not harm anything other than the organisms they were designed to control.

Pollution by Heavy Metals

The atomic mass and density of heavy metals are higher as compared to other elements. There are more than 20 heavy metals, but only a few of them, such as cadmium (Cd), copper (Cu), argon (Ar), silver (Ag), chromium (Cr), zinc (Zn), lead (Pb), uranium (Ur), radium (Ra), nickel (Ni), *etc.*, are considered, due to their toxicity. Heavy metals are hazardous metals. All food and water contain metals and nonmetals that, at high concentrations, can become harmful. Inorganic fertilizer contains heavy metals as impurities, rock phosphate, being a high potential source, always implies the addition of a significant amount of lead and cadmium into the soil. A combination of low analysis and straight fertilizers may add more lead and cadmium to the soil.

The heavy metals occur in solution as cations and are absorbed by the negatively charged soil particles. Heavy metal pollution is serious because it can persist for many decades.

Pollution by Organic Waste

The pollution potential of organic waste (urban and rural) has become a national and international problem, especially concerning soil pollution. Because organic waste contains garbage, sewage effluents, and agricultural waste, it causes serious soil pollution. It is necessary to ascertain the level of inorganic and organic chemicals in the sewage sludge before its application to the cropland.

Microorganisms have the potential to recover pollutants or decompose them. Biodegradation can completely mineralize a xenobiotic compound, resulting in a clean-up or partial degradation of products that are subsequently detoxified by abiotic mechanisms, spontaneous breakdown, or biotically by another group of microbes. Biotransformation without biodegradation can ameliorate the toxicity associated with organic pollutants *via* humification or polymerization. Biostabilization reduces the risks associated with the leaching of organic matter from the soil. Application of bioremediation for the clean-up of soil sediments, groundwater, or industrial waste streams is an urgent priority for soil and environmental microbiologists.

BIODEGRADATION

Biodegradation refers to the degradation of organic contaminants in the soil as well as groundwater by indigenous or transplanted/acclimated microorganisms, primarily bacteria and fungi. Organic pollutants are converted into carbon dioxide, water, and microbial cell mass under aerobic conditions (in the presence of oxygen). Under anaerobic conditions (in the absence of oxygen), methane, small quantities of carbon dioxide and hydrogen, and occasionally intermediate species, which may be less equally or more hazardous than the metabolized compound, are produced. There are some methods of biodegradation, which are as follows:

Ex Situ Method

This method is applicable to the remediation of polluted soil. The contaminated soil can be remediated by excavation, detoxification, or destruction of physical and chemical contaminants. Under the processes of stabilization, solidification, and immobilization, contaminated soil undergoes un-contamination.

In Situ Method

Destruction or transformation of the soil contaminants is mainly carried out through the *in situ* method. The physico-chemical methods for soil remediation render plant growth and remove all biological activities and microorganisms such as N_2 fixing bacteria, PSB, soil fungi, and fauna [3].

PHYTOREMEDIATION

It is a remediation method where a plant is used for the removal or detoxification of contaminants like soil, water, air, *etc.* [4]. Phytoremediation includes several techniques such as phytostabilization, rhizofiltration, phytoextraction, phytovolatilization, and phytotransformation [5].

Phytostabilization

Phytostabilization means reducing the mobility of contaminants like heavy metals in soil. Plants are grown in the field to stabilize and reduce the bioavailability of pollutants by binding them with soil particles, making them less available to plants and humans [6].

Rhizofiltration

It is a process that filters water through roots or absorbs pollutants from contaminated soil or solution. Rhizofiltration is often used to remove contaminants from groundwater directly. In this process, translocation and tolerance are highly irrelevant [6].

Phytoextraction

Phytoextraction is also known as phytoaccumulation or phytosequestration. Plants are grown to exploit the contaminated soil and harvested plant biomass. Plant roots take up substances from the environment, where pollutant content is very high and accumulates in their aerial parts. Phytoaccumulation can be done by populous, salix, *Brassica* spp., *etc.*

Presently, several methods are used for the extraction or removal of contaminants from the environment. It includes chemical precipitation, ion exchange, electrochemical treatment, adsorption on activated carbon, *etc.* [7]. Among them, chemical precipitation and electrochemical treatment methods are more effective for decontamination, while the heavy metal ion concentration is low. Ion exchange and activated carbon adsorption processes are extremely expensive compared to other methods. The removal of pollutants from the environment through bioagents has been directed towards bioremediation [8].

The biodegradation of pesticides is often complex and involves a series of biochemical reactions. Several enzymes efficiently catalyze the biodegradation of pesticides. The full understanding of the biodegradation pathway often requires new investigations.

MICROBIAL DEGRADATION

Degradation of Organochloride Pesticides Through Soil Microbes

Microorganisms are more important in the degradation of pesticides than physicochemical mechanisms. The major biochemical reactions involved in pesticide breakdown are oxidation, reduction, hydrolysis, and synthetic reactions.

Some chlorinated pesticides, such as dichloro-diphenyl-trichloroethylene (DDT) and dieldrin, have been shown to be resistant. Consequently, they remain in the environment for a longer period of time and accumulate in the food chains for several years after their application to the soil system [9]. The pure culture method is used for the biodegradation of chlorinated pesticides in most of the studies. Usually, the culture is isolated from the soil sample, which is contaminated with chlorinated pesticides. The isolated strains are characterized, classified, and tested with several concentrations of the pesticide. DDT-metabolizing microorganisms have been isolated from different habitats, including soil, water, animal faces, sewage, activated sludge, marine, and sediments of freshwater [10]. Heavy inoculation of DDT degrading microbes, *e.g.*, *Aerobacter aerogenes* in soil, has been shown to cause the disappearance of DDT from the soil system. The degradation of chlorinated pesticides by pure cultures has been proven. Matsumura and co-workers [11] reported that *Pseudomonas* spp. can degrade dieldrin in the soil. Several microorganisms can degrade the selected organochlorine pesticides present in soil (Table **1**).

Table 1. Degradation of organochloride pesticides by soil microbes.

S.No.	Organochloride Pesticides	Microbes Responsible for Degradation
1	PCP	*Arthrobacter* sp.
		Flavobacterium sp.
2	1,4- Dichlorobenzene	*Pseudomonas* sp.

(Table 1) cont.....

S.No.	Organochloride Pesticides	Microbes Responsible for Degradation
3	DDT	*Aerobacter aerogenes*
		Trichoderma viride
		Pseudomonas sp.
		Micrococcus sp.
		Arthrobacter sp.
		Bacillus sp.
4	Lindane	*Sphingomonas paucimobilis*
		Streptomyces sp.
		Pleurotus ostreatus
5	DDE	*Phanerochaete chrysosporium*
6	DDD	*Trichoderma* sp.
7	Heptachlor epoxide	*Phlebia* sp.
		Phanerochaete sp.
8	Aldrin	*Micrococcus* sp.
		Bacillus sp.
		Trichoderma viride
9	Dieldrin	*Pseudomonas* sp.

Degradation of Organophosphate Pesticides through Soil Microbes

The most popular organophosphate pesticide in India is methyl parathion [(O, O-dimethyl-O-(p-nitro-phenyl phosphorothioate)], which is used on a number of crops. It is widely used throughout the world and its residues are regularly detected in different vegetable and fruit crops. Research is investigating the degradation of these pesticides through microorganisms is useful. For this, several bacteria have been isolated for the degradation of methyl parathion worldwide [12, 13]. Of those, *Pseudomonas putida* and *Acinetobacter rhizosphaerae* bacteria can degrade the organophosphate pesticides rapidly.

Degradation of Carbamate Pesticides Through Soil Microbes

It has been investigated that the microbial degradation of carbamates can be done in carbamate pesticides. In most cases, the research studies did not eliminate the possibility that abiotic processes are involved in the degradation. Carbofuran is one of the popular pesticides that belongs to the N-methylcarbamate class, which is extensively used in agriculture. It exhibits high toxicity to mammals and is highly hazardous. Several bacteria, including *Pseudomonas* spp., *Flavobacterium*, *Achromobacter*, *Sphingomonas*, and *Arthrobacter* spp., are capable of degrading carbofuran, and their characterization is being done to better understand the bacterial role in removing residual carbofuran from the environment. Carbofuran is first degraded into carbofuran phenol and then into 2-hydroxy-3-(3-methylpropan-2-ol) phenol by *Sphingomonas* sp [14]. and *Arthrobacter* sp [15], respectively.

BIOREMEDIATION

Bioremediation is a process used to detoxify or remove organic and inorganic pollutants from the environment. Bioremediation means the use of microorganisms or plants, non-viable or viable, natural or genetically engineered, to treat environments polluted with organic molecules that are problematic to break down (xenobiotics) and to remove toxic heavy metals by altering them into little or no toxic elements, henceforth forming innocuous products [16]. This method is effective only when the environmental conditions are favourable for the growth and development of microbes [17]. In the past, wastes from agricultural, domestic, and industrial sources were regularly treated by different conventional methods, such as chemical precipitation, electrochemical treatment, and ion exchange. These methods supply only some degree of active treatment for them, and they are very expensive. Bioremediation is an active and economical method for the treatment of toxic metals and other wastes.

Bioremediation of Heavy Metals

Heavy metal pollution in the environment is receiving great attention worldwide. A number of toxic substances, including heavy metals such as arsenic (As), zinc (Zn), cadmium (Cd), uranium (U), selenium (Se), mercury (Hg), lead (Pb), chromium (Cr), silver (Ag), gold (Au) and nickel (Ni), are very hazardous to the environment. In addition to that, the presence of these pollutants, *i.e.*, heavy metals, in the environment has been a hot topic due to their toxicity, non-biodegradability, and almost indefinite persistence in the environment. Among the living entities, microorganisms play a great role in the bioremediation of heavy metals originating from contaminated soil, industrial waste, and wastewater. When microbes, especially bacteria, are exposed to a higher concentration of heavy metal, it may have cidal effects on them. Microorganisms have developed diverse approaches to overcome the toxic effects of heavy metals and metalloids present in contaminated soil through their utilization and accumulation by reducing their toxicity or bio-availability through such processes, *i.e.*, biomethylation and transformation. Microorganisms may interact with pollutants such as heavy metals and radionuclides through several mechanisms, some of which may be used as the basis for potential bioremediation strategies [18]. Microorganisms act on heavy metals through some mechanisms, which are biosorption (sorption of metal to the cell surface by a physicochemical method), bioleaching (removal of heavy metals through excretion of organic acids or methylation reactions), bio-mineralization (immobilization of metals through the formation of an insoluble compound or polymeric complex), intracellular accumulation, and enzyme-catalyzed transformation [19]. The variability of microorganisms such as bacteria, fungi, yeasts, and algae has been studied for use in bioremediation processes, and some of these have been employed for the bioremediation of heavy metals (Table 2). Bioremediation is cost-effective and highly efficient compared to conventional methods for heavy metal mitigation.

Table 2. A list of metal degrading microorganisms.

S.No.	Metals	Degrading Microorganisms
1.	Cr	*Pseudomonas aeruginosa, Bacillus subtilis*, and *Saccharomyces cerevisiae*
2.	Cd	*Alcaligenes* sp., *Pseudomonas* sp., and *Moraxella* sp.
3.	Ni	*Bacillus subtilis* and *P. licheniformis*
4.	Ag	*Streptomyces noursei*
5.	Au	*Aspergillus niger* and *Chlorella pyrenoidosa*
6.	Co	*S. cerevisiae*
7.	Cu	*Candida tropicalis* and *Bacillus licheniformis*
8.	Fe	*Bacillus subtilis*
9.	Hg	*Penicillium chrysogenum*
10.	Mn	*Bacillus licheniformis*
11.	Pb	*P. chrysogenum*
12.	U	*S. cerevisiae*
13.	Th	*S. cerevisiae*
14.	Zn	*Rhizopus arrhizus, Penicillium chrysogenum*, and *P. spinulosum*

Bioremediation of Agricultural Waste

Approximately 38 billion metric tonnes of organic waste are produced every year by humans, livestock, and crops worldwide. Disposable and eco-friendly management of these organic wastes has become a global priority. Therefore, more attention has been given in recent years to developing cost-effective and efficient technologies to convert agricultural waste or organic waste into value-added products for sustainable development. However, these wastes can be managed through different methods, such as vermicomposting, vermiwash, and some microorganisms. Vermicomposting is nothing but organic manure, where earthworms change agricultural waste or organic waste into nutritional organic manure. Here, microorganisms, especially soil bacteria, degrade the organic matter, and earthworms release some beneficial compounds like hormones, organic acid, and enzymes into the substrate, which is called vermicompost. Several species of earthworm, such as *Eisenia fetida*, *Perionyx excavatus*, *P. sansibaricus*, and *Eudrilus eugeniae*, have been identified as organic waste feeders (detritus) and can be used potentially to minimize the ill effects of anthropogenic waste, which is received from different sources. Agricultural byproducts such as crop residue, litter, animal dung, poultry waste, *etc.* are a rich source of nutrients for plants. According to conservation estimation, about 6 to 7 hundred million tonnes of agricultural waste are available in India. This huge quantity of organic waste can be converted into organic manure, organic pesticides, and organic liquid fertilizer by earthworms, beneficial microorganisms, and biofertilizers. Vermicompost is a nutrient-rich organic manure, which often results in mass reduction, less time for preparation, and reduced phytotoxicity in ready material. Several combinations of organic waste and animal dung can be used for vermicomposting to obtain value-added products like enriched vermicompost, enriched FYM, compost, *etc.* These organic products, rich in carbon and plant nutrients, indicate a potential for using agricultural wastes in sustainable crop production.

Limitation of Bioremediation

Biodegradation has some common limitations, which are related to hazardous chemical waste containing high waste concentration and toxicity. Sometimes, this toxicity either inhibits the growth and development of microorganisms or kills them. For the optimum growth of microorganisms, it is necessary to have a sufficient amount of moisture, mineral nutrients and a favourable soil pH. They require a favourable temperature for their proper growth, which ranges from 20-30°C. Once the limitations of bioremediation are corrected, the ubiquitous distribution of microorganisms will be there to allow them to carry out their major responsibility. In the majority of cases, an inoculation with specific microorganisms is neither necessary nor useful in most cases. Besides all these factors, the nature and chemical composition of waste, the solubility of the pollutant, the oxidation–reduction potential of waste, and microbial interaction also affect the bioremediation processes.

CONCLUDING REMARKS

Microbes can significantly affect the degradation of organic and inorganic pollutants present in the soil system since they have developed the means to use them for their own benefit. Biological degradation by micro-organisms can efficiently remove pesticides from the environment as well as the food chain, especially chlorinated, organophosphates, and carbamates used in agriculture. It is concluded that bioremediation techniques have been successfully employed in the field and are gaining more importance with the increased acceptance of environmentally friendly solutions.

CONSENT FOR PUBLICATION

Not applicable.

CONFLICT OF INTEREST

The author declares no conflict of interest, financial or otherwise.

ACKNOWLEDGEMENTS

Declared none.

REFERENCES

[1]   Meenambigai P, Vijayaraghavan R, Gowri RS, *et al.* Biodegradation of heavy metals-a review. Int J Curr Microbiol Appl Sci 2016; 5: 375-83.

[2]   Garima T, Singh SP. Application of bioremediation on solid waste management: a review. J Bioremediat Biodegrad 2014; 5: 248. [doi: 10.4172/2155-6199.1000248].

[3]   Burns RG, Rogers S, McGhee I. Remediation of Inorganics and Organics in Industrial and Urban Contaminated Soils. In: Naidu R, Kookana RS, Oliver DP, Eds. *et al.* Contamination and the Soil Environment in the Australasia-Pacific Region. Netherlands: Springer 1996; pp. 411-49.

[4]   Palmroth MR, Pichtel J, Puhakka JA. Phytoremediation of subarctic soil contaminated with diesel fuel. Bioresour Technol 2002; 84: 221-8.
        [PMID: 12118697]

[5]   Ghosh M, Singh SP. A review on phytoremediation of heavy metals and utilization of it's by products. Asian J Energy Environ 2005; 6: 214-31.

[6]   Latha MR, Indirani R, Kamaraj S. Bioremediation of polluted soil. Agric Rev (Karnal) 2004; 25: 252-66.

[7]   Matheickal JT, Yu Q, Woodburn GM. Biosorption of cadmium (II) from aqueous solutions by pre-treated biomass of marine alga *Durvillaea potatorum.* Water Res 1999; 33: 335-43.

[8]   Das N, Vimala R, Karthika P. Biosorption of heavy metals – an overview. Indian J Biotechnol 2008; 7: 159-69.

[9]   Kannan K, Tanabe S, Williams RJ, *et al.* Persistent organochlorine residues in foodstuffs from Australia, Papua New Guinea and the Solomon Islands: contamination levels and human dietary exposure. Sci Total Environ 1994; 153: 29-49.
        [PMID: 7939620]

[10]   Lal R, Saxena DM. Accumulation, metabolism, and effects of organochlorine insecticides on microorganisms. Microbiol Rev 1982; 46: 95-127.
        [PMID: 6178010]

[11]   Matsumura F, Boush GM, Tai A. Breakdown of dieldrin in the soil by a micro-organism. Nature 1968; 219: 965-7.

[PMID: 5673021]

[12]   Liu Z, Hong Q, Xu JH, *et al.* Cloning, analysis and fusion expression of methyl parathion hydrolase. Acta Genet Sin 2003; 30: 1020-6.

[13]   Liu H, Zhang JJ, Wang SJ, *et al.* Plasmid-borne catabolism of methyl parathion and *p*-nitrophenol in *Pseudomonas* sp. strain WBC-3. Biochem Biophys Res Commun 2005; 334: 1107-14.
       [PMID: 16039612]

[14]   Kim IS, Ryu JY, Hur HG, *et al. Sphingomonas* sp. strain SB5 degrades carbofuran to a new metabolite by hydrolysis at the furanyl ring. J Agric Food Chem 2004; 52: 2309-14.
       [PMID: 15080638]

[15]   De Schrijver A, De Mot R. Degradation of pesticides by actinomycetes. Crit Rev Microbiol 1999; 25: 85-119.
       [PMID: 10405795]

[16]   Gupta A, Joia J, Sood A, *et al.* Microbes as potential tool for remediation of heavy metals: a review. J Microb Biochem Technol 2016; 8: 364-72.

[17]   Vidali M. Bioremediation- an overview. Pure Appl Chem 2001; 73: 1163-72.

[18]   Lloyd JR, Anderson RT, Macaskie LE. Bioremediation of Metals and Radionuclides. Atlas R, Philp J, Eds. Bioremediation: Applied Microbial Solutions for real-World Environmental Cleanup. Washington: ASM Press 2005; pp. 293-317.

[19]   Lloyd JR. Bioremediation of metals: the application of microorganisms that make and break minerals. Microbiol Today 2002; 29: 67-9.

CHAPTER 9

Soil Microbe-Mediated Bioremediation: Mechanisms and Applications in Soil Science

Anandkumar Naorem[1,*], **Shiva Kumar Udayana**[2], **Jaison Maverick**[3] and **Sachin Patel**[1]

[1] *Indian Council of Agricultural Research-Central Arid Zone Research Institute, Regional Research Station-Bhuj, Gujarat-370105, India*

[2] *Krishi Vigyan Kendra, Dr. YSRHU, Venkataramannagudem, West Godavari, Andhra Pradesh-534101, India*

[3] *Department of Agricultural chemistry and Soil Science, Bidhan Chandra Krishi Viswa Vidyalaya, Mohanpur, West Bengal-741252, India*

Abstract: Bioremediation is a prominent and novel technology among decontamination studies because of its economic practicability, enhanced proficiency, and environmental friendliness. The continuously deteriorating environment due to pollutants was taken care of by the use of various sustainable microbial processes. It is a process that uses microorganisms like bacteria and fungi, green plants, or their enzymes to restore the natural environment altered by contaminants to its native condition. Contaminant compounds are altered by microorganisms through reactions that come off as a part of their metabolic processes. Bioremediation technologies can be generally classified as *in situ* or *ex situ*. *In situ* bioremediation involves treating the pollutants at the site, while *ex situ* bioremediation involves the elimination of the pollutants to be treated elsewhere. This chapter deals with several aspects, such as the detailed description of bioremediation, factors of bioremediation, the role of microorganisms in bioremediation, different microbial processes and mechanisms involved in the remediation of contaminants by microorganisms, and types of bioremediation technologies such as bioventing, land farming, bioreactors, composting, bioaugmentation, biofiltration, and bio-stimulation.

INTRODUCTION

The increased rate of industrialization and the rigorous release of toxic chemicals containing high amounts of heavy metals have led to serious environmental contamination. Consequently, a contaminated environment poses a huge concern for human health, and in addition to that, the increasing population and rapid urbanization necessitate the amelioration of polluted sites for restoration and

---

[*] **Corresponding author Anandkumar Naorem:** Indian Council of Agricultural Research-Central Arid Zone Research Institute, Regional Research Station, Bhuj, Gujarat-370105, India; E-mail: naoremanand@gmail.com

Ashutosh Gupta, Shampi Jain & Neeraj Verma (Eds.)

economic use. The most inconvenient, time-consuming, and laborious method of remediation is simply digging up the polluted soil and dumping it into a landfill. It does not ensure the remediation of the polluted soil; rather, it transports polluted soil from one place to another, presenting an appreciable risk as a result of the excavation, handling, and shifting of hazardous materials.

Among various techniques, the use of living organisms like plants (phytoremediation) and microbes (bioremediation) or their activities to degrade or decontaminate environmental pollutants is an economical and ecologically safe means to detoxify the contaminated environment and is increasingly used as a substitute to expensive common heavy metal remediation methods such as soil flushing and thermal vitrification. The bioremediation process involves the biological degradation of organic wastes in a regulated environment to a safe level or levels below the critical limits provided by concerned authorities. Bioremediation makes use of living organisms, especially soil microorganisms, to decontaminate soil and water pollutants into less toxic states. It exploits innate soil microorganisms or plants to reduce or eliminate the risk of heavy metal entry into the food chain. Microorganisms use several processes and mechanisms to decontaminate polluted soil or water. These heavy metal-loving microbes may either belong to the native contaminated soil or be isolated from different soils and later inoculated into the contaminated soil.

Heavy metals are being used by soil microbes for their energy and metabolic activities for proper growth and development. These soil microbes have multifaceted properties that help in the biodegradation of a compound. The process in which specific beneficial microorganisms are inoculated into a contaminated site and allowed to reproduce and act on these contaminants is called bioaugmentation. For bioremediation to be efficacious, microbes, with the help of their enzymes, break down the contaminants and transform them into less toxic substances. It can be successfully employed provided there are optimum environmental conditions for the growth and development of beneficial soil microbes. Therefore, its execution often necessitates the control of soil, water, and air variables to permit better microbial growth and further breakdown of contaminants to progress at a rapid rate.

A few heavy metals play crucial roles in the growth and development of living organisms [1], but at higher concentrations, they may result in the formation of unspecific compounds, leading to cytotoxic effects [2]. It is acknowledged that the heavy metal availability in soil significantly affects the microbial community and structure through alterations in several biochemical processes, especially the fungal and bacterial community and diversity [3]. Some researchers report that bacteria are resistant to heavy metals [4], which could be due to biochemical processes that can detoxify heavy metal toxicity [5] and/or release chelating agents [6] that form complexes with heavy metals and reduce their phytotoxicity. For example, some common chelating agents can form soluble complexes with lead, gold, and uranium in plants, leading to a surge in the metal solubility resulting in increased transfer from root parts to shoot portions [7].

Shenker and coworkers [8] indicated that phytosiderophores enhance iron absorption and facilitate the transfer of other ions within the plant. Both bacteria and fungi produce a significant number of siderophores [9]. These siderophores differ based on the type of microorganism and prevailing environmental conditions. The chemical properties of these siderophores are different for each microbe, which allows them to survive in different types of environments [10].

ROLE OF MICROORGANISMS IN BIOREMEDIATION

Microorganisms of discrete groups can break down various pollutants relative to their concentration and their metabolic demand. Microbes play a major role and have immense importance in bioremediation in different types of ecosystems. These microorganisms break down pollutants by producing certain types of enzymes.

To classify a soil microorganism under bioremediation, it must meet certain criteria [11]:

a. The microorganism must be able to survive and signal its biological activity in contaminated environments.
b. The microorganisms must be able to reach contaminants that are commonly found adsorbed to soil particles.
c. The microorganism must be capable of producing powerful enzymes that are responsible for bioremediation.
d. These enzymes' active sites must be accessible to the substrate.
e. The enzymatic and accompanying systems must have close intracellular or extracellular contact.
f. Microbes must be able to survive and thrive in harsh conditions.

One of the most commonly found soil microorganisms related to bioremediation is *Pseudomonas* species. It is interesting to note that *Pseudomonas putida* was the first reported microbial patent (1974) registered for its petroleum degrading ability. Petroleum and oil slicks have also been reported to be biodegraded by *Corynebacterium*, *Mycobacterium*, and pseudomonads [12]. Numerous aerobic bacteria are also known to degrade pesticides and hydrocarbons (both alkanes and polyaromatic), which often include *Pseudomonas*, *Alcaligenes*, *Sphingomonas*, *Rhodococcus*, and *Mycobacterium*.

There is growing attention to the utilization of anaerobic bacteria in the bioremediation of polychlorinated biphenyls (PCBs) in river sediments and the dichlorination of the solvents trichloroethylene (TCE) and chloroform. Anaerobic bacteria are known to decontaminate PCBs, chloroform, dechlorinate, and TCE in the sediments of lakes and river beds. For instance, ligninolytic degrading fungi like *Phanerochaete chrysosporium* are effective in the bioremediation of recalcitrant poisonous environmental pollutants such as plastics, TNT, DDT, and numerous high molecular weight polynuclear aromatics. Methylotrophs can bioremediate contaminants like 1,2-dichloroethane and chlorinated aliphatic trichloroethylene. *Geobacter metallireducens* has the capability to diminish metals and transform them into non-toxic or less toxic forms, especially in the removal of uranium from mining operations, water draining systems, and also groundwater [13]. A detailed mechanism of the pollutant-microbe interaction that leads to the cleaning up of the pollutants in the soil is shown in Fig. (**1**).

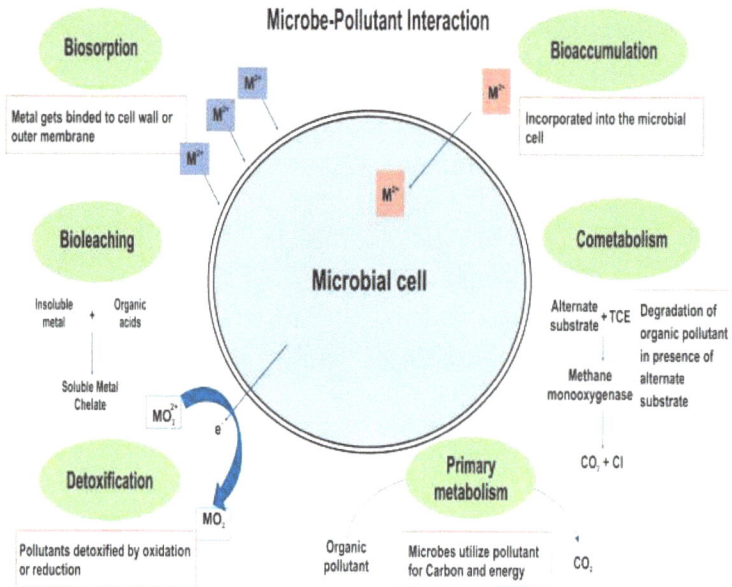

Fig. (1). Detailed mechanisms of microbe-pollutant interaction.

Numerous hazardous compounds that are dismissed as being mineralized by a single microbe can be decontaminated by using a consortium of such beneficial microbes. Among these hazardous compounds, 2, 4, 6-trinitrotoluene (TNT), polychlorinated biphenyls, polyaromatic hydrocarbons, aliphatic and aromatic halogenated organics, *etc*., have been identified as being degraded by microbial consortia. Microbial consortiums feed on these pollutants in a sequential process. This contaminant/pollutant needs to be in contact with the bacteria for degradation, and this is not certainly attained as neither the microbes nor contaminants occur homogenously in the soil. A few bacteria are mobile and manifest a chemotactic response, sensing the contaminant and advancing towards it.

ENVIRONMENTAL VARIABLES AFFECTING BIOREMEDIATION

Several factors influence the rate, type, and efficiency of bioremediation. These factors can be broadly categorized into three types: contaminant, microbial, and environmental factors (Fig. **2**).

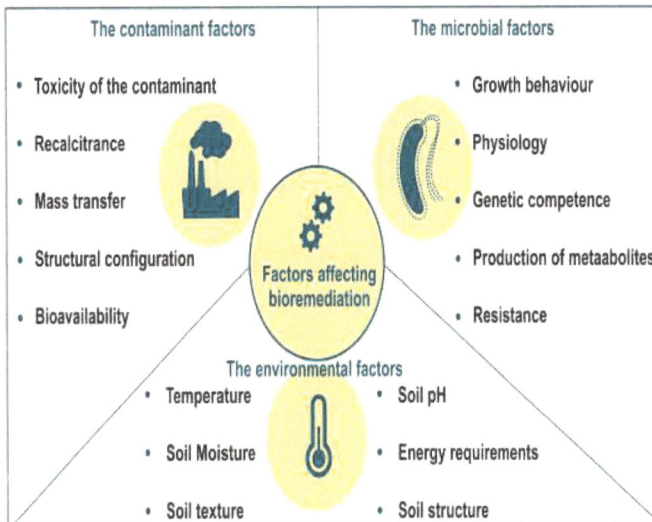

Fig. (2). Factors affecting the process of bioremediation.

Concentration of Pollutant

Microbes' interactions with substrate have likely effects on the degradation rates of distinctive contaminants [14]. The substrate interaction assessment at varying concentrations is imperative due to its major role in bacterial sensitivity and metabolism. The synergistic interactions between different components of a contaminant can accelerate the decomposition rates and catabolic enzyme activities. For example, the growth rate of *Pseudomonas putida* in a batch culture study was found to be suppressed at high substrate concentrations [15]. At particular concentrations, the benzene, toluene, ethylbenzene, and xylene (BTEX) compounds exhibited inhibitory effects on microbiological processes due to their complex interactions [16].

Acclimatization (Microbial Adaptation)

Acclimatization is the practice in which a specific organism adapts to a change in its environment, enabling it to uphold performance across a range of environmental conditions. The adaptation of microbial communities to aromatic organic compounds augments their breakdown efficiency [17]. For example, *Alcaligenes xylosoxidans* Y234 degraded benzene and toluene more efficiently under acclimatized conditions in contrast to non-adapted ones [18].

Bioavailability of Carbon

The prime factor ascertaining the biodegradation rate of organic contaminants to microbes is approachability [19]. The bioavailability of hydrocarbons is a function of their physical attributes and chemical composition [20]. Some bacteria obtain substances that act as additives and proliferate the overall degradation of petroleum hydrocarbons [21]. For example, biosurfactants released by innumerable microbes like bacteria and fungi have the potential to contribute to the uptake and mineralization of petroleum hydrocarbons [22].

Nutritional Requirements

Despite the presence of microbes in the polluted soil, their population is not as large as required for bioremediation. Their growth and activity must be intensified. Microorganisms necessitate bio-stimulation, which generally entails the addition of nutrients and oxygen. These nutrients permit the microbes to produce important enzymes to detoxify the pollutants. Carbon is the most basic element of living forms and is required in higher quantities than other elements. Carbon, hydrogen, oxygen, and nitrogen together account for 95% of the weight of cells. Phosphorous and sulphur account for about 70% of the remainder. The nutritional requirement with respect to carbon to nitrogen is 10:1, whereas carbon to phosphorous is 30:1.

ENVIRONMENTAL REQUIREMENTS FOR BIOREMEDIATION

Temperature

Temperature plays an important role in microbial growth and development and also in hydrocarbon degradation [23]. A reduction in microbial growth and augmentation is noticed at low temperatures, resulting in a decreased rate of petrochemical degradation [24]. Diverse studies have indicated an increased solubility of hydrocarbons and easy accessibility of petrochemical hydrocarbons to microorganisms with enhanced temperatures. When it comes to the degradation of petroleum wastes, marine environments require a lower temperature (15–20°C) than soil (30–40°C) [25]. However, some heat-resistant microbes (*e.g. Bacillus thermoleovorans*) have been reported to effectively alter phenanthrene, naphthalene, and anthracene even at elevated temperature levels.

pH

The changes in pH principally affect the bioremediation of various contaminants in soil. Several research findings suggest that most of the organic pollutants can be ameliorated at around pH 7. A modest change in pH has a considerable effect on overall bioremediation in a marine ecosystem. However, under exceedingly acidic conditions, fungi and acidophilic microbes are reported to grow and perform the biodegradation process [26].

Oxygen Availability

Aerobic or anaerobic environmental conditions are mainly determined by the presence of oxygen. Aerobic environments are the most favourable conditions for the degradation of petrochemical hydrocarbons, and in a few cases, they may also be degraded in anaerobic environments in aquifers and submerged marine sediments. Aquifers and submerged marine sediments are the sites where anaerobic biodegradation prevails the most [27], with an insignificant degradation rate and is primarily limited to halogenated aromatics. The activity of mono- and dioxygenases is dependent on oxygen for aerobic petrochemical degradation.

MECHANISMS INVOLVED IN BIOREMEDIATION

Microorganisms have a wide range of processes and mechanisms for transforming chemicals in their surroundings. In certain cases, the contaminants serve as carbon and energy sources for microbial growth, while they act as terminal electron acceptors in others. This diverse ability of microbes inspires the use of microbes to modify and detoxify contaminant molecules. Numerous steps and details of the microorganisms' actions are described below.

Primary Substrate Utilization

Primary substrate utilization occurs when a microbe mutually alters a substrate and utilizes it as an energy or carbon source. An electron acceptor is the main requisite for these transformations. It can be either aerobic or anaerobic, even though the presence of oxygen favours the reactions. This type of biodegradation is efficient in the breakdown of petroleum compounds and a few pesticides.

Co-Metabolism (Secondary Substrate Utilization)

A few microbes do not exploit the target contaminant for their growth but rely on alternative compounds for their growth and survival–a process known as co-metabolism. For example, methane monooxygenase, an enzyme released during the oxidation of methane by methanotrophic bacteria, can alter chlorinated solvents such as TCE. During the degradation of the substance, it seems that the organism entailed has no net carbon or energy gain and may even bring about a by-product that is lethal to the cell. For the enzymatic activity to take place at a rapid pace and also for the conversion of TCE, methanotrophic bacteria require an organic growth substrate like propane or butane.

Reductive and Hydrolytic Dehalogenation

The principal components of pesticides and hazardous industrial wastes are chlorides and other halogens, and separating them from the pesticides may result in a drop in the toxicity of the pesticides. If the halogen is substituted by a single hydrogen (RCl -> RH), it is reductive dehalogenation. If two halogens are substituted concomitantly, then the process is called dihaloelimination, although it still comes under reductive dehalogenation. If the halogen is displaced by OH (RCl -> ROH), then it is hydrolytic dehalogenation. In both cases, the halogen is liberated in its inorganic form into the environment.

Detoxification and Activation

Detoxification involves the transformation of harmful contaminant molecules into innocuous substances. It is done by processes like hydrolysis, hydroxylation, dehalogenation, demethylation, methylation, and ether cleavage. By breaking bonds, or adding or removing groups, the organism reduces the toxic effect of chemicals on the environment.

Microorganisms use the product as a carbon source and modify its chemical nature until it is released as CO_2. In some instances, the initial compound may be harmless, but the substance produced by microorganisms, or an intermediate in the degradation process, may become a toxin. This process is called activation. For this reason, it is important to test all steps of a reaction when determining how a compound is degrading.

The new toxins may also be more or less mobile than their predecessors, so they can either stick around in one area for extended periods or spread to other areas and cause damage. A prevalent example of this is the dechlorination of TCE, which produces DCE (50 times more hazardous than TCE) and vinyl chloride (a known carcinogen). Commonly used insecticides in the past, like zinophos, trichloronat, and carbofuran, were all found to increase the soil's toxicity with extended use.

TYPES OF BIOREMEDIATION

There are two broad categories of bioremediation, *viz.*, *in situ* and *ex situ*. The type of remediation depends on the nature of the contaminant and the microorganism used.

In Situ Remediation

In-situ remediation technology involves the remediation of contaminated material in field conditions. Bioaugmentation, bioventing, and bio-stimulation are some of the technologies of bioremediation. *In situ* bioremediation depends on the activity of native microorganisms in field conditions to detoxify and degrade the contaminants. The activity of native microbes in turn depends on the availability of substrate, nutrients, and terminal electron acceptors.

Biosorption

Biosorption is an *in situ* technology based on metal and ion adsorption from soil or aqueous solution by microorganisms. It involves the binding of metals to the cell walls of microbes. Kim and coworkers [28] used *Trichoderma reesei* to detoxify cadmium and copper ions *via* adsorption and desorption. Sari and Tuzen [29] have used *Amanita rubescens* as a biosorbent for detoxifying Pb (II) and Cd (II).

Bioaugmentation

Bioaugmentation is the introduction of specific microbes into a polluted environment that can degrade particular contaminants or contaminant groups. The microbes may be naturally occurring or genetically engineered to target a contaminant. It is done by isolating and culturing the microorganisms, followed by the introduction of such microbes into contaminated sites for remediation purposes. Bioaugmented microbial communities that are isolated from benzene, toluene, ethylbenzene, and xylene (BTEX) sites are found to be very efficient in bioremediation of BTEX-contaminated environments. Generally, bioaugment ation using many taxa of microorganisms, *viz.*, as a microbial consortium (*e.g.*, microbial mats and assemblages of bacteria–microalgae/ cyanobacteria), has been found to be more efficient.

Applications

PAH degradation is up to 90% when remediated with bioaugmentation [30]. *Dehalococcoides* sp., which dechlorinates trichloroethylene (TCE) to ethene, was found to be effective for bioaugmentation application on TCE contaminated sites. The development of encapsulated bacteria inocula has shown significant results in the remediation of pentachlorophenol-contaminated sites [31]. PAHs and naph-thenic acids are remediated using bioaugmenting microbes *viz.Fusibacter*, *Alkaliphilus*, *Desulfobacterium*, *Variovorax*, *Thauera*, and *Hydrogenophaga* in anaerobic biodegradation [32].

Limitations

The characterization of the site and microorganisms needs proper and extensive study. In some cases, contaminants that are solubilized are prone to contaminating groundwater by leaching.

Bioventing

It is a modern technology where an external oxygen supply is provided to stimulate and enhance the native microbes for the remediation of contaminants. The biodegradation is mainly aerobic and *in situ*. This technology is used for the degradation of BTEX compounds in contaminated aquifers by stimulating aerobic degradation. The airflow rates are maintained low to supply the necessary oxygen to microbes. It is done by direct air injection into the contaminated soils. Apart from BTEX compounds, some volatile contaminants are also degraded by bioventing.

Applicability

Alkane and PAHs are degraded by bioventing using microorganisms like *Alcanivorax*, *Marinobacter*, *Thalassospira*, *Alteromonas*, and *Oleibacter* [33]. Suko and coworkers [34] reported that volatilization was followed by biodegradation of petroleum hydrocarbons by the application of bioventing. Sui and coworkers [35] used bioventing for the decontamination of pesticides and a few organic pollutants.

Limitations

The major limitations of bioventing are: (1) low permeability of soil, (2) vapour build-up within the radius of influence of air injection wells, and (3) low moisture and temperature, which reduce the degradation rate significantly.

Bioleaching

Bioleaching is the use of microorganisms that can solubilize specific pollutants in water that can be removed by leaching. Bioleaching is mainly used to recover metals from the environment. Microorganisms like *Thiobacillus ferrooxidans* and *T. thiooxidans* solubilize copper, silver, zinc, and uranium. Solubilization is mainly done by the substitution of protons with absorbed metals or by the oxidation of particular metals, making them soluble.

Application

Kumar and Philip [36] reported that bioleaching is a very efficient way to remove and dispose of heavy metals from. White and coworkers [37] used sulfur-oxidizing bacteria to leach metals such as Cd, Co, Cr, Mn, Ni, and Zn through sulfuric acid production and precipitation. Mishra and coworkers [38] showed the use of *Acidithiobacillus ferrooxidans* in the solubilization of lithium ion from secondary batteries.

Co-Metabolic Process

The co-metabolic process is mainly used for the remediation of non-petroleum hydrocarbons. It mainly involves the production of enzymes and the degradation of other compounds. The main requirement of co-metabolism is the availability of a suitable substrate that aids in the metabolism of the target contaminant. In some cases, water containing oxygen and methane is applied to groundwater to enhance methanotrophic microorganisms for degrading chlorinated hydrocarbons.

Limitation

Co-metabolism will not be effective in a heterogeneous subsurface layer. Circulation of methane solutions will be difficult when there is heterogeneity in the contaminated zone.

Monitored Natural Attenuation

Monitored natural attenuation (MNA) is the reliance on natural processes for remediation of contaminant pollutants. MNA requires evaluation and modelling of contaminant degradation, sensitive receptor impact, exposure pathways, and downgradient of contaminant concentration. MNA is a multidisciplinary process where effectiveness is usually considered on a case-by-case basis.

Applicability

MNA is successful in the remediation of fuel hydrocarbons. Fuel and halogenated volatile organic compounds are generally evaluated for MNA. Oil degradation for a longer duration is found to be promoted by natural attenuation. Grčman and coworkers [39] have used EDTA (ethylenediaminetetraacetic acid) along with some plants to increase the extraction of heavy metals like cadmium, lead, and zinc from soil.

Limitations

(1) Modelling using site-specific data is required; (2) products after degradation may be mobile and toxic; (3) remediation takes longer than other remediation measures; (4) geochemical and hydrologic conditions change over time, causing remobilization of the stabilized contaminant; and (5) long-term monitoring and associated costs.

Biostimulation

Biostimulation is the process of the addition of nutrients that stimulate the growth and activity of native microorganisms, which increases their metabolic activity, thus elevating the degradation.

Application

Margesin and Schinner [40] have found that biostimulation of microorganisms by the application of nutrients at 10°C to wastewater effectively decontaminates anionic surfactants and fuel oil. They used biostimulation with fertilizer (N-P-K) for decontaminating diesel oil spills and found a 22% higher decontamination in fertilized soils compared to unfertilized soils.

Biofilter

A biofilter is the use of microorganisms like bacteria to act as filters of contaminants from polluted water and waste. Wastewater contaminated by metal ions like Cu^{2+} and Cd^{2+}, silver nanoparticles, and trichloroethylene [41] can be remediated using a biofilter.

Applications

Methanotrophic bacterium *Methylosinus trichosporium* is used in biofilters for the decontamination of water contaminated by trichloroethylene [41]. A micro-algal/bacterial biofilter is used in the detoxification of Cu and Cd metal wastes [42]. Duran and coworkers [43] used a bacterium, *Chromobacterium violaceum*, to treat an effluent contaminated by silver nanoparticles, effectively reducing its concentration.

Ex Situ **Remediation**

Ex situ treatment technologies involve the physical removal of contaminated material from its original place and treatment in another area. Biopiles, composting, and land farming are examples of *ex situ* technologies.

Biopiles

Biopiles is the process where soils are removed and mixed with soil amendments and placed in enclosures with aeration and leachate collection systems. Biopiles are used for the remediation of petroleum hydrocarbon contaminated soils. The treatment site is lined with impermeable liners to minimize the leaching of contaminants into groundwater. The addition of nutrients enhances microbial degradation. Soil piles are raised 2-3 m in height, while these can be raised as high as 20 ft. To maintain positive pressure, an air distribution system is placed below the soil. Runoff, evaporation, and volatilization are controlled by covering piles with plastic sheets. The remediation period in biopiles will be from a few weeks to several months. In the presence of volatile organic compounds (VOC), discharging air should be treated to remove contaminants before it escapes into the atmosphere.

Applicability

Biopiles are generally used for the remediation of non-halogenated VOCs and fuel hydrocarbons. However, the results varied and may not be applicable for certain compounds.

Limitations

Soil excavation and management, application of nutrients and oxygen, and periodic testing are the major limitations of biopile treatments.

Composting

Composting is a regulated microbe-mediated process in which the contaminated soil is assorted with bulking agents and organic amendments like wood chips, hay, manure, green waste, *etc.* in an appropriate quantity to supply a balanced carbon and nitrogen supply essential for the activities of thermophilic microbes. Organic pollutants like polycyclic aromatic hydrocarbons (PAHs) and 1,1,1-trichloro-2,2-bis (p-chlorophenyl) ethane (DDT) are transformed into less harmful products by microbial activity in the presence or absence of oxygen. The heat emitted by the native microbes during the decomposition process results in a thermophilic phase (55-65°C) during the composting process, and this phase is of prime importance for proper conversion of the harmful pollutants. The effectiveness of the decomposition process can be achieved by maintaining the optimum oxygen content (by windrow turning), moisture content (by irrigation), and temperature.

Distinct layouts and designs are employed in composting. These are (1) aerated static piles, wherein compost piles are aerated through blowers or vacuum pumps; (2) in-vessel composting with mechanical agitation, wherein compost is positioned in a reactor vessel and assorted and aerated; and (3) windrow composting, a more economical method, in which compost is positioned in long piles called windrows and intermittently stirred using mobile equipment.

Applicability

Biodegradable pollutants present in the contaminated soils and sediments can be remediated using this process. A few pilot and full-scale projects have been started that use aerobic and thermophilic composting to reduce the concentration of explosives such as trinitrotoluene (TNT), RDX, HMX, and ammonium picrate to acceptable levels. By this method, PAHs and DDT residues can also be degraded.

Limitations

The main drawbacks of composting are (1) the requirement of considerable space, (2) soil excavation is desired, and (3) a large amount of additive materials is needed.

Land Farming

Land farming is a prevalent bioremediation technology wherein the excavated contaminated soils are positioned onto lined beds of a fixed thickness and aerated intermittently by turning the soils. For the degradation process to proceed at an ideal rate, moisture content, aeration, pH, nutrients, and soil amendments need to be supplied at a regulated rate. Microbial decomposition is favoured under aerobic conditions by increasing the accessibility of nutrients and moisture in this process.

Applicability

Land farming has been effective in degrading petroleum hydrocarbons while the chlorinated or nitrated compounds are treated rather slowly. It is also effective in treating diesel fuel, fuel oils, oil sludge, wood preservation wastes, coke wastes, and pesticides.

Limitations

Land farming has limitations like (1) necessity for a large area; (2) rainfall and temperature may not be regulated, which hamper the multiplication and growth of microbes, leading to increased degradation periods; (3) volatile compounds and gases may be released to the atmosphere, which should be taken care of; (4)

requirement for constructing a provision to assemble and supervise the runoff waste; and (5) site characterization for relief, erosion, climate, permeability, *etc.*, for proper design of the facility.

Other Types

Fungal Remediation

Fungal metabolism is found to be convenient in the successful treatment of various organic contaminants, especially hydrocarbons. The white-rot fungus (*Phanerochaete chrysosporium*) has the capability to break down an immense range of organic pollutants, including PCBs, PAHs, and explosives. The white-rot fungus produces lignin peroxidases, which are accountable for extensive degradation ability.

Applications

The fungus *Fusarium oxysporum* can detoxify and mineralize discrete concentrations of anthracene, phenanthrene, and pyrene in soil. On the other hand, the white-rot fungus *Phanerochaete chrysosporium* can detoxify chlorinated hydrocarbons, PAHs, PCBs, polychlorinated (p) dioxins, pesticides (lindane and DDT), and some azo dyes.

Limitations

The restrictions for fungal remediation are (1) their high sensitivity to biological processes; (2) their inability to thrive effectively in suspension systems; (3) the negative impact of mixing on the production of enzymes; (4) the inability of fungus to sequester tightly to fixed media; (5) toxicity; (6) chemical sorption; (7) competition with native microbes; and (8) sluggish modification ability.

Microalgal/Cyanobacterial Remediation

Microalgae and cyanobacteria can be sourced to ascertain pollution and also decontaminate various pollutants in the surroundings. Microalgae and cyanobacteria may assist as bioindicators of pollution in long-term insecticide-contaminated soils. The manifestation of cyanobacterial mats in the Arabian Gulf Coasts after oil pollution has been acknowledged enough due to their effectiveness in degrading hydrocarbons.

CONCLUDING REMARKS

Bioremediation is a promising technology that can be employed alone or in combination with common remediation alternatives for the comprehensive management of a diverse group of contaminants in the soil. Bioremediation is a reliable approach for contamination degradation, and it is more economical and eco-friendly than other technologies that are often used to decontaminate hazardous waste. Microorganisms play a very important role in the bioremediation of different types of environmental pollutants. Therefore, there is a need to enhance our knowledge of microbial structure and function and their potential for environmental decontamination, which leads us towards the insight of microbial genetics in enhancing the pollutant detoxifying potential of effective organisms. Future aspects of the research should be directed towards the knowledge of the interactions of pollutants with the environment using multidisciplinary technologies, which will help us to choose the most sustainable method for the remediation of pollutants from the environment.

CONSENT FOR PUBLICATION

Not applicable.

CONFLICT OF INTEREST

The author declares no conflict of interest, financial or otherwise.

ACKNOWLEDGEMENTS

Declared none.

REFERENCES

[1]   Higgins IJ, Burns RG. Eds. The Chemistry and Microbiology of Pollution. London: Academic Press Inc 1975; p. 248.

[2]   Nies DH. Microbial heavy-metal resistance. Appl Microbiol Biotechnol 1999; 51: 730-50.
        [PMID: 10422221]

[3]   Rajapaksha RMCP, Tobor-Kapłon MA, Bååth E. Metal toxicity affects fungal and bacterial activities in soil differently. Appl Environ Microbiol 2004; 70: 2966-73.
        [PMID: 15128558]

[4]   Silver S, Phung LT. Bacterial heavy metal resistance: new surprises. Annu Rev Microbiol 1996; 50: 753-89.
        [PMID: 8905098]

[5]   Gadd GM, Griffiths AJ. Microorganisms and heavy metal toxicity. Microb Ecol 1977; 4: 303-17.
        [PMID: 24232222]

[6]   Nowack B. Chelating agents and the environment. Environ Pollut 2008; 153: 1-2.
        [PMID: 18336972]

[7]   McGrath SP, Zhao FJ, Lombi E. Plant and rhizosphere processes involved in phytoremediation of metal-contaminated soils. Plant Soil 2001; 232: 207-14.

[8]   Shenker M, Fan TWM, Crowley DE. Phytosiderophores influence on cadmium mobilization and uptake by wheat and barley plants. J Environ Qual 2001; 30: 2091-8.
        [PMID: 11790018]

[9] Prahbu V, Biolchini PF, Boyer GL. Detection and identification of ferricrocin produced by ectendomycorrhizal fungi in the genus *Wilcoxina*. Biometals 1996; 9: 229-34.

[10] Winkelmann G. Ecology of siderophores with special reference to the fungi. Biometals 2007; 20: 379-92.
[PMID: 17235665]

[11] Alexander M, Ed. Biodegradation and Bioremediation. San Diego, USA: Academic Press Inc. 1994; p. 302.

[12] Sardrood BP, Goltapeh EM, Varma A. An Introduction to Bioremediation. Goltapeh EM, Danesh Y, Varma A, Eds. Fungi as Bioremediators. Berlin: Springer-Verlag 2013; pp. 3-27.

[13] Kumar A, Bisht B, Joshi V, Dhewa T. Review on bioremediation of polluted environment: a management tool. Int J Environ Sci 2011; 1: 1079-93.

[14] Wang L, Barrington S, Kim JW. Biodegradation of pentyl amine and aniline from petrochemical wastewater. J Environ Manage 2007; 83: 191-7.
[PMID: 16678965]

[15] Abuhamed T, Bayraktar E, Mehmetoglu T, *et al.* Kinetics model for growth of *Pseudomonas putida* F1 during benzene, toluene and phenol biodegradation. Process Biochem 2004; 39: 983-8.

[16] Mathur AK, Majumder CB. Kinetics modelling of the biodegradation of benzene, toluene and phenol as single substrate and mixed substrate by using *Pseudomonas putida*. Chem Biochem Eng Q 2010; 24: 101-9.

[17] Babaarslan C, Abuhamed T, Mehmetoglu U, *et al.* Biodegradation of BTEX compounds by a mixed culture obtained from petroleum formation water. Energy Sources 2003; 25: 733-42.

[18] Yeom SH, Kim SH, Yoo YJ, *et al.* Microbial adaptation in the degradation of phenol by *Alcaligenes xylosoxidans* Y234. Korean J Chem Eng 1997; 14: 37-40.

[19] Guthrie EA, Pfaender FK. Reduced pyrene bioavailability in microbially active soils. Environ Sci Technol 1998; 32: 501-8.

[20] Foght JM, Westlake DWS, Johnson WM, *et al.* Environmental gasoline-utilizing isolates and clinical isolates of *Pseudomonas aeruginosa* are taxonomically indistinguishable by chemotaxonomic and molecular techniques. Microbiology 1996; 142: 2333-40.
[PMID: 8828201]

[21] Lee PH, Ong SK, Golchin J, *et al.* Use of solvents to enhance PAH biodegradation of coal tar-contaminated soils. Water Res 2001; 35: 3941-9.
[PMID: 12230177]

[22] Leahy JG, Colwell RR. Microbial degradation of hydrocarbons in the environment. Microbiol Rev 1990; 54: 305-15.
[PMID: 2215423]

[23] Margesin R, Schinner F. Biodegradation and bioremediation of hydrocarbons in extreme environments. Appl Microbiol Biotechnol 2001; 56: 650-63.
[PMID: 11601610]

[24] Gibb A, Chu A, Wong RCK, *et al.* Bioremediation kinetics of crude oil at 5°C. J Environ Eng 2001; 127: 818-24.

[25] Mueller JG, Chapman PJ, Pritchard PH. Creosote-contaminated sites: their potential for bioremediation. Environ Sci Technol 1989; 23: 1197-201.

[26] Stapleton RD, Savage DC, Sayler GS, *et al.* Biodegradation of aromatic hydrocarbons in an extremely acidic environment. Appl Environ Microbiol 1998; 64: 4180-4.
[PMID: 9797263]

[27] Coates JD, Anderson RT, Lovley DR. Oxidation of polycyclic aromatic hydrocarbons under sulphate

reducing conditions. Appl Environ Microbiol 1996; 62: 1099-101.
[PMID: 16535261]

[28]   Kim SK, Park CB, Koo YM, *et al.* Biosorption of cadmium and copper ions by *Trichoderma reesei* RUT C30. J Ind Eng Chem 2003; 9: 403-6.

[29]   Sari A, Tuzen M. Kinetic and equilibrium studies of biosorption of Pb(II) and Cd(II) from aqueous solution by macrofungus (*Amanita rubescens*) biomass. J Hazard Mater 2009; 164: 1004-11.
[PMID: 18845395]

[30]   Mishra D, Kim DJ, Ralph DE, *et al.* Bioleaching of metals from spent lithium ion secondary batteries using *Acidithiobacillus ferrooxidans.* Waste Manag 2008; 28: 333-8.
[PMID: 17376665]

[31]   Stormo KE, Crawford RL. Preparation of encapsulated microbial cells for environmental applications. Appl Environ Microbiol 1992; 58: 727-30.
[PMID: 16348656]

[32]   Folwell BD, McGenity TJ, Price A, *et al.* Exploring the capacity for anaerobic biodegradation of polycyclic aromatic hydrocarbons and naphthenic acids by microbes from oil-sands-process-affected waters. Int Biodeter Biodegr 2016; 108: 214-21.

[33]   Catania V, Santisi S, Signa G, *et al.* Intrinsic bioremediation potential of a chronically polluted marine coastal area. Mar Pollut Bull 2015; 99: 138-49.
[PMID: 26248825]

[34]   Suko T, Fujikawa T, Miyazaki T. Transport phenomena of volatile solute in soil during bioventing technology. J ASTM Int 2006; 3: 1-6.

[35]   Sui H, Li X, Jiang B, *et al.* Simulation of remediation of multiple organic contaminants system by bioventing. *HuagongXuebao/* J Chem Indus Engine (China) 2007; 58: 1025-31.

[36]   Kumar M, Philip L. Bioremediation of endosulfan contaminated soil and water - optimization of operating conditions in laboratory scale reactors. J Hazard Mater 2006; 136: 354-64.
[PMID: 16730891]

[37]   White C, Sharman AK, Gadd GM. An integrated microbial process for the bioremediation of soil contaminated with toxic metals. Nat Biotechnol 1998; 16: 572-5.
[PMID: 9624690]

[38]   Mishra S, Jyot J, Kuhad RC, *et al.* Evaluation of inoculum addition to stimulate *in situ* bioremediation of oily-sludge-contaminated soil. Appl Environ Microbiol 2001; 67: 1675-81.
[PMID: 11282620]

[39]   Grčman H, Velikonja-Bolta S, Vodnik D, *et al.* EDTA enhanced heavy metal phytoextraction: metal accumulation, leaching and toxicity. Plant Soil 2001; 235: 105-14.

[40]   Margesin R, Schinner F. Bioremediation (natural attenuation and biostimulation) of diesel-oil-contaminated soil in an alpine glacier skiing area. Appl Environ Microbiol 2001; 67: 3127-33.
[PMID: 11425732]

[41]   Taylor RT, Hanna ML, Shah NN, *et al. In situ* bioremediation of trichloroethylene-contaminated water by a resting-cell methanotrophic microbial filter. Hydrol Sci J 1993; 38: 323-42.

[42]   Loutseti S, Danielidis DB, Economou-Amilli A, *et al.* The application of a micro-algal/bacterial biofilter for the detoxification of copper and cadmium metal wastes. Bioresour Technol 2009; 100: 2099-105.
[PMID: 19109013]

[43]   Duran N, Marcato PD, De Souza GIH, *et al.* Antibacterial effect of silver nanoparticles produced by fungal process on textile fabrics and their effluent treatment. J Biomed Nanotechnol 2007; 3: 203-8.

CHAPTER 10

Role of Soil Microbes in Sustainable Development: Nutrient Transformation, Bioremediation, and Biodeterioration

Anurag Singh[1], **Shreya Kapoor**[1], **Priya Bhatia**[1], **Sanjay Gupta**[2], **Nidhi S. Chandra**[1] and **Vandana Gupta**[1,*]

[1] *Department of Microbiology, Ram Lal Anand College, University of Delhi, Benito Juarez Road, New Delhi-110021, India*

[2] *Independent Scholar (Ex Head and Professor, Department of Biotechnology, Jaypee Institute of Information Technology), Sector 62, Noida, UP-201309, India*

Abstract: Pedogenesis, or the formation of soil, takes decades along with a combination of parent geological material, natural biota, distinct climate, and topography. Soil, which hosts rich functional biodiversity ranging from microbes to higher plants, provides nutrients, anchorage for roots, holds water, and buffers against pollutants. After going through this chapter, readers will be able to appreciate how nature takes care of the nutritional requirements of its dwellers, how these nutrients, in turn, get transformed following the life-death cycle, and the infallible role that soil microbes play in this process. We aim to describe how the enormous but bio-unavailable nutrient sources, both in the atmosphere (nitrogen) and the earth's crust (phosphorus, iron, *etc.*), are made accessible to plants in a multi-step mechanism. Curiosity and concern among mankind have provoked a wide range of scientific developments. Nevertheless, exploitative anthropogenic activities have degraded this vital life-supporting component. All kinds of pollutants and unsustainable agricultural practices over time have deposited harmful and toxic chemicals in the soil, the negative effects of which are being deliberated lately. Soil microbes hold promise in remediating these xenobiotic compounds and providing economically feasible and ecologically safe solutions. In the final section, we provide a brief overview of the ability of microbes to utilize a range of substrates that can prove detrimental to both modern infrastructure and archaeological artifacts.

Keywords: Bioaugmentation, Carbon cycle, Nitrogen fixation, Wood rotting, Wood staining.

---

* **Corresponding author Vandana Gupta:** Department of Microbiology, Ram Lal Anand College, University of Delhi, Benito Juarez Road, New Delhi-110021, India; E-mail: vandanagupta72@rediffmail.com

Ashutosh Gupta, Shampi Jain & Neeraj Verma (Eds.)

INTRODUCTION

Soil is a rich ecosystem consisting of its own biotic and abiotic components. It has a large diversity of living organisms; plants, nematodes, earthworms, and microorganisms, including bacteria, fungi, actinomycetes, protozoa, *etc.*, which are known to interact among themselves and maintain the soil's microclimate. Curiosity and concern for mankind, followed by scientific development, have led to a vast array of discoveries that have facilitated knowledge buildup. Nitrogen was discovered in 1772 by Daniel Rutherford, which further paved the way for other scientists to unwrap its importance in biological systems, its presence in decaying matters, and ultimately the complete cycling of nitrogen in nature [1].

Advancements in science have allowed us to delineate the vital role that microorganisms play in the cycling of enormous but bio-unavailable elemental nitrogen to its reduced forms, which plants can utilize, and again replenishing it back into the atmospheric sink. Second, like nitrogen, phosphorus also makes up an essential requirement for soil fertility. However, it is seldom bioavailable owing to its highly reactive characteristics and presence in the soil as either insoluble salts or organic phosphate. Doran and Safley [2] defined soil health as *its capacity to function as a vital living system, to sustain biological productivity, promote the quality of air and water environments, and maintain plant, animal, and human health.* The soil microorganisms carry out nutrient mineralization and assimilation [3]. These can be utilized by plants either immediately (mineralized nutrients) or following the death of microbes (assimilated nutrients). In this chapter, we will discuss the presence of microbes that facilitate nutrient transformations and contribute to soil health and other ecological functions.

Talking of microorganisms, one ought to appreciate their role as ultimate decomposers. They maintain the continuity of nutrient cycles by breaking down dead living matter in a process called biodegradation. This essentially involves the transformation of complex compounds into simpler, generally less toxic forms that can be easily taken up by the environment and then cycled. According to the principle of microbial infallibility, all-natural substances are degradable under suitable conditions [4]. However, microorganisms fail to degrade most synthetic compounds (xenobiotics) as their chemical structure and bonding are not recognized by the degradative enzymes found among the autochthonous community. Also, the buildup of natural compounds such as crude oil in large quantities in the environment beyond the degradative ability of the autochthonous community leads to the accumulation of these pollutants in the ecosystem, especially the soil. In the following sections, we will discuss how bioremediation provides an economically feasible and ecologically safe solution to this problem. The presence of natural metabolic capabilities in certain soil bacteria or secondary metabolites produced by others allows their use in the degradation or detoxification of many toxic pollutants. For example, while *Rhodococcus* can metabolize pesticides, fungi like *Phanerochaete chrysosporium* can secrete extracellular enzymes to digest them [5, 6].

It is now well established that microbes can inhabit even the extremes of habitats owing to their ability to utilize a wide range of substrates. This adaptability can also be credited to their presence as mixed cultures in biofilms on the surface, where pioneer residents facilitate subsequent successions. In the final section of this chapter, we discuss how this characteristic of microbes, coupled with essential phenomena of biodegradation, can wreak havoc and elicit economic losses in a process called biodeterioration. Iron, which today finds its use in various sectors, is vulnerable to corrosion. Moreover, microbes like *Thiobacillus* and *Desulfovibrio* further augment this process. Microbiologically induced corrosion (MIC) can occur under both aerobic and anaerobic conditions. Biodeteriogens can even act on surfaces like stones, concrete, and polyethylene by secreting exoenzymes, acids, noxious compounds, and chelating agents, among others.

Soil Microbes in Nutrient Transformation

Nutrient transformation is the conversion of unavailable forms of nutrients through various processes such as mineralization (conversion of organics to minerals, thus inorganic forms) and assimilation (conversion of minerals to organic forms) to a more bioavailable form. Soil microorganisms are vital to this process and thus have a great impact on soil health and other organisms dwelling in soil ecosystems. Nitrogen and phosphorus are two important but limiting nutrients for plant growth. Although abundant in nature, they remain unavailable to plants. However, the transformations facilitated by microorganisms make these essential nutrients accessible to them.

Soil Microbes in Nitrogen Transformation

Nitrogen is abundant in the atmosphere. However, this gaseous nitrogen cannot be directly utilized by the plants and needs to be reduced to a more usable form (ammonia) by a process known as biological nitrogen fixation (BNF) [7, 8]. Soil microorganisms mediate the transformation of gaseous nitrogen into usable reduced form ammonia. The other major bioavailable form of nitrate, which is generated by the process of nitrification of ammonia, is further converted back to elemental nitrogen in a complete nitrogen cycle (Fig. **1**), the steps of which are discussed below [9, 10].

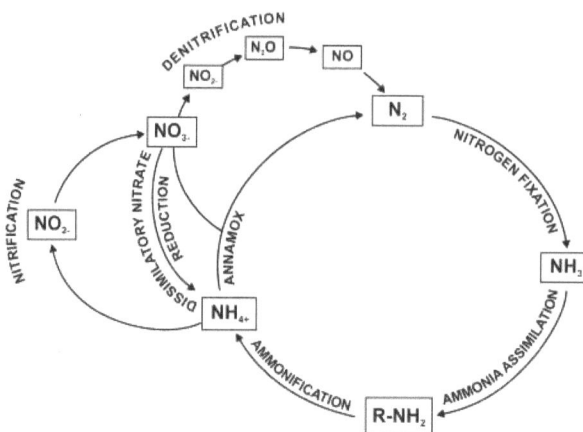

Fig. (1). Nitrogen transformation in the soil.

Nitrogen Fixation

It is an energy-expensive process involving the conversion of elemental nitrogen into organic forms facilitated by a group of microorganisms referred to as biological nitrogen fixers. Among nitrogen fixers, the bacterial population is called diazotrophs, which can either be free-living (*e.g.*, *Azotobacter, Beijerinckia*) [11] or symbiotic (*e.g.*, *Rhizobium, Frankia*) [12]. These organisms fix nitrogen by employing an oxygen-sensitive nitrogenase enzyme complex [12 - 14]. The reduced form of nitrogen is then assimilated to form nitrogenous bases of nucleic acid, amino acids, and proteins. Further, the reduced nitrogen in amino acids needs to be mineralized for the flow of nutrients. Many microorganisms like *Bacillus*, *Pseudomonas*, *Flavobacterium*, and *Clostridium* convert these organic nitrogenous compounds into ammonia or ammonium ions by ammonification, which can then be assimilated by plants. This process can take place extracellularly or intracellularly in both aerobic and anaerobic conditions. Ammonia (or ammonium ions) can also be oxidized in a chemolithotrophic two-step [8] energy-yielding nitrification process to yield nitrite and then nitrate ions. Ammonium oxidation is catalyzed by ammonium mono-oxygenase [9], and the energy yielded is then used by microorganisms for carbon metabolism [8, 9, 13]. Nitrification is a pH-sensitive process (optimal range of 6.6-8.0) that gets slowed down as the pH drops and can be carried out by both autotrophic and heterotrophic bacteria. *Nitrosomonas* spp. (*N. eutropha* and *N. marina*), *Nitrosospira* spp., *Nitrosolobus* spp. and *Nitrosovibrio* spp. carry out ammonia oxidation, while *Nitrobacter winogradskyi*, *N. vulgaris*, *Nitrospina gracilis*,

Nitrococcus mobilis, etc., facilitate nitrite oxidation. The presence of both types of nitrifiers in close proximity prevents the accumulation of nitrite. The nitrate ions thus produced can then be utilized by a variety of organisms by either assimilatory or dissimilatory nitrate reduction. In the former case, several enzymes, *e.g.*, nitrate reductase, are employed by organisms like *Thiobacillus*, *Rhizobium japonicum*, and *Azotobacter* [15] to convert nitrate to ammonia and successively to amino acids. In the absence of oxygen, organisms like *Clostridium*, *Citrobacter*, *Enterobacter*, and *Erwinia* use nitrate as a terminal electron acceptor while oxidizing organic substrates [13]. Assimilatory nitrate reduction is sensitive to the presence of ammonia but insensitive to oxygen, while dissimilatory nitrate reduction is inhibited by the presence of oxygen but remains insensitive to ammonia. Moreover, nitrate can also be completely reduced to elemental nitrogen *via* nitrite (NO_2^-), nitric oxide (NO), and nitrous oxide (N_2O) in an anaerobic (or low oxygen conditions) process known as denitrification [10]. *Pseudomonas*, *Alcaligenes*, *Rhodopseudomonas*, *Propionibacterium*, *Bacillus*, and *Flavobacterium* can carry out denitrification (Fig. **1**).

Anammox is an acronym for anoxic ammonia oxidation, a chemolithotrophic process [8], which utilizes nitrite as a terminal electron acceptor to reduce ammonium ions to dinitrogen [9]. Bacteria belonging to genera like *Brocadia*, *Kuenenia*, *Scalindua*, *etc.* can facilitate this reaction [9, 16]. Although predominant in an aquatic habitat, their presence has also been reported in certain soil types with an anoxic condition (wetland soils, permafrost soils, and agricultural soils) [16]. These bacteria thrive in aerobic-anaerobic interfaces to utilize both ammonium and nitrite ions to carry out the anammox reaction [9].

Soil Microbes in Phosphorus Transformation

Phosphorus is an essential macronutrient governing plant growth and nutrition. It is required for various processes, *viz.*, the formation of seeds, root development, shoot development, photosynthesis, fixation of nitrogen, *etc.* Phosphorus can be found at the soil surface in either organic or insoluble inorganic forms. The soil organic phosphorus is further divided into three categories depending upon the bonds present: phosphate esters (consisting of ester linkages) make up phosphoproteins, nucleic acids, phospholipids, *etc.*; phosphoric acid anhydrides (having anhydride bonds in addition to the monoester linkages) like ATPs, ADPs, and phosphonates (having carbon-phosphorus linkages) serve as the components of phosphonoglycans and other extracytoplasmic macromolecules [17, 18]. On the other hand, inorganic phosphorus constitutes 70-80% of the total phosphorus found in the soil and is present in iron, aluminium, or calcium complexes, *e.g.* tricalcium phosphate [19]. These insoluble forms are not readily accessible by plants, making phosphorus a limiting growth factor [20].

A large number of microorganisms referred to as phosphate-solubilizing microorganisms (PSMs) mediate phosphate solubilization (conversion of insoluble inorganic phosphorus into its soluble form) and mineralization (conversion of insoluble organic phosphorus into its soluble form) [20, 21].

Phosphate solubilization occurs by several mechanisms: Firstly, reduction in soil pH, which is achieved by the production of organic acids like gluconic acid (*e.g. Erwinia herbicola*), 2-ketogluconic acid (*e.g. Bacillus firmus*), citric acid (*e.g. Azospirillum* sp.), tartaric acid (*e.g. Bacillus* sp.), succinic acid (*e.g. Pseudomonas* sp.), oxalic acid (*e.g. Aspergillus*), and CO_2 that dissolves in soil water to produce carbonic acid [21, 22], or by cation release during ammonia assimilation. Secondly, PSMs like *Penicillium rugulosum* secrete inorganic acids such as nitric acid and HCl that facilitate the conversion of triphosphates (*e.g.* TCP) to di or monophosphates.

Enzymes like phytases (myo-inositol hexakisphosphate) and phosphatases mediate phosphorus mineralization. The former facilitates hydrolysis of phytate, the main source of organic phosphorus found in seeds, pollen, *etc* [20, 23], while the latter is known for carrying out dephosphorylation of phosphoanhydride or phospho-ester bonds in organic compounds like sugar phosphates [20]. *Colletotrichum graminicola* and *Aspergillus niger* are known for the production of acid phosphatases and phytases, respectively. Further, phosphonatases and C–P lyases facilitate phosphate mineralization by mediating the cleavage of organophosphates. Factors like temperature, climate, soil type, and the interactions of PSMs with other microbes influence the solubilization and mineralization processes.

Future Aspects

Even though an array of soil microorganisms mediating nutrient transformations has been delineated, we still look to agrochemicals to enhance agricultural production, which has its repercussions. This knowledge gap in identifying functional strains of microbes and their respective mechanisms of action, along with the impact of different environmental conditions and human interventions on their activity, needs to be worked on. Further research on these factors and regulating transformations could generate more reliable evidence of their significant role in ecosystem functioning and their holistic application in sustainable agricultural practices.

SOIL MICROBES IN BIODEGRADATION AND BIOREMEDIATION

A key role is played by microorganisms in the degradation of detritus, *i.e.*, the dead and decaying plant matter as well as the carcasses and waste products of animals. Plants are important sources of carbon, so the degradation of plant polymers such as cellulose, hemicellulose, and lignin is necessary to replenish the soil to carry on the cycling of carbon. Higher organisms such as earthworms and basidiomycetes fragment the detritus, which increases the surface area for the activity of these microorganisms. Bacteria and fungi derive nutrients for growth from these substrates directly or indirectly through co-metabolism. Larger molecular weight compounds are converted into smaller products that can be taken up by the cells. This is achieved by the secretion of extracellular enzymes such as cellulases, ligninolytic, and proteolytic enzymes.

Process of Biodegradation

Biodegradation is a series of processes involving the action of a diverse range of microbial enzymes on artificial and natural compounds to convert them into less, similarly harmful, or more harmful intermediates than their parental compounds. In the case of aerobic conditions, the intermediates are eventually converted to readily soluble inorganic compounds, carbon dioxide, and water (complete oxidation). Whereas, under anaerobic conditions, the organic compounds are partially oxidized into simple organic acids, methane, and hydrogen gas. Some intermediates are not acted upon by microbial enzymes and stay in the environment as recalcitrant.

Factors affecting the rate of the biodegradation process are soil moisture, pH, oxygen, nutrient availability, temperature, the concentration of contaminants, and the presence of appropriate microbes.

Generally, there are three steps in the process:

1. Biodeterioration: It is the negative aspect of biodegradation as modifications in the physical, mechanical, and chemical properties of the useful substrates are a result of microbial growth that leads to economic losses such as the deterioration of textiles, wool, paper, leather, *etc.*

2. Bio-fragmentation: It refers to the breakdown of polymer chains into monomers or oligomers by the action of microbial enzymes.

3. Assimilation: Uptake of nutrients, carbon, and energy sources from fragmented polymers by the microorganisms and then transformation into biomass, water, and carbon dioxide.

Principles of Bioremediation

Bioremediation is the use of living organisms or their metabolites to minimize or eliminate environmental pollution caused by potentially toxic, or recalcitrant compounds. Bioremediation is a great alternative to conventional methods of remediation, as complete mineralization of pollutants can be accomplished at a lower cost [4]. The basic principles of bioremediation involve either complete mineralization or reducing the solubility of these environmental contaminants by changing pH, or adsorption of contaminants from polluted environments [24].

How do Microbes Destroy Contaminants?

The use of bacteria for the degradation and detoxification of numerous toxic chemicals, such as pesticides, crude oil, *etc.*, is an effective tool to decontaminate polluted sites. Isolation of indigenous bacteria capable of metabolizing pollutants provides an environment-friendly means of *in situ* detoxification [25]. The biodegradative capability of microbes is dependent on many environmental factors such as temperature, pH, water potential, nutrients, and the concentration of pollutants in soil, which may limit the growth of degrading microorganisms [26]. Various plasmids in soil bacteria contain genes encoding degradative enzymes [27, 28]. *Pseudomonas, Alcaligenes, Actinobacteria, Flavobacterium, Klebsiella, Moraxella*, and *Arthrobacter* are some commonly found genera of soil bacteria with catabolic plasmids [29]. Some of these bacteria, *e.g.*, *Rhodococcus*, are reported to have the ability to metabolize pesticides [30 - 32].

Non-pathogenic bacteria residing in the rhizosphere or as endophytes in internal plant tissues can also degrade toxic organic compounds. *Acetobacter, Arthrobacter, Bacillus, Burkholderia, Enterobacter, Herbaspirillum*, and *Pseudomonas* are some examples of endophytic bacteria found in the soil [33]. These bacteria produce a variety of metabolites that aid in plant growth and the biodegradation of soil pollutants [34, 35]. Consortia of bacteria are preferred in mineralizing xenobiotic pesticides over single isolates as co-metabolism has been shown to enhance the rate of pesticide biodegradation.

White-rot fungi like *Phanerochaete chrysosporium* can degrade a broad spectrum of recalcitrant organic compounds. They secrete many extracellular enzymes like peroxidase or ligninolytic enzymes, mainly lignin peroxidase (LiP), manganese peroxidase (MnP), and laccase, which not only allow them to digest cross-linked phenolic polymers like lignin but many complex aromatic bonds of pesticides and dyes too. Polycyclic aromatic hydrocarbons (PAHs) such as pyrene, anthracene, di- and tri benzoic acids, several polychlorinated biphenyls (PCBs), DDT (dichlorodiphenyltrichloroethane), lindane, and chlordane, for example, have all been shown to degrade in the presence of this non-specific enzyme system [4, 36].

Trametes sp., *Polyporus* sp., *Nigroporus* sp., *Fusarium solani* F33, and *Helminthosporium carbonum* U11 are some other biodegradative fungi shown to degrade 2, 8-DCDD (2, 8-dichlorodibenzo-p-dioxin) and DDT by secreting extracellular peroxidases [37].

The Role of Microorganisms in Bioremediation

The simplest form of bioremediation is natural attenuation (NA) or bioattenuation, which employs the natural metabolic capabilities of indigenous microbial populations to degrade recalcitrant compounds [38, 39]. This process is slow, and the indigenous populations may not be able to degrade the pollutants. One of the possible reasons for this could be that their degradative plasmids lack the genes that would encode for a suitable degradative enzyme. Even if the autochthonous microorganisms can degrade the pollutant, keeping their numbers high is a challenge as the nutrients available at the site and the energy generated from the degradation are hardly enough for their sustenance [40]. In the absence of other more efficient remedial techniques, natural attenuation can be used to treat sites with a lower concentration of environmental pollutants [38].

The obstacle of nutrient scarcity at the polluted site can be overcome by the addition of easily metabolized nutrients such as glucose and also by the addition of limiting nutrients such as nitrogen and phosphorus. This process is termed biostimulation [41]. The use of glucose as a co-substrate was shown to hasten dicofol biodegradation by isolated rhizosphere-associated microorganisms [42]. While the addition of nutrients did increase bacterial activity, this method was found to be costly and cumbersome for larger contaminated sites.

Bioaugmentation is used when the aforementioned techniques fail to effectively decontaminate the polluted site. It is defined as the introduction of wild-type or genetically modified microbes to enhance the rate of biodegradation of recalcitrant pollutants [38]. The microbe that is added to the polluted soil is known as a bio-inoculant, and it can be introduced to the soil through various methods, like the bacterization of plant seeds. Rahman and co-workers [6] proposed that co-inoculation of a consortium of bacteria, each with a unique catabolic degradation mechanism, is more efficient than the inoculation of one individual strain with a complete catabolic pathway. *Alcaligenes, Arthrobacter, Bacillus, Brucella, Burkholderia, Catellibacterium, Pichia, Pseudomonas, Rhodococcus, Serratia, Sphingomonas, Stenotrophomonas, Streptomyces,* and *Verticillium* are some bacterial genera that have been reported to be useful in the bioremediation of pesticide-contaminated soils using bioaugmentation technology. Other microorganisms that have been used to carry out bioremediation are listed in Table **1**.

This technique also has its own set of limitations. Firstly, the number of introduced microorganisms depletes with time as a result of abiotic and biotic stresses. The abiotic factors include fluctuations in pH, temperature, moisture content, nutrient availability, as well as possibly lethal concentrations of the pollutant. The biotic stresses are a result of the competition of allochthonous microorganisms with indigenous populations for nutrients, along with antagonistic interactions such as predation by protozoa, bacteriophages and fungi, and amensalism due to antimicrobial-producing populations [43]. Secondly, the bioinoculant is generally unable to penetrate less porous layers to reach the depths of the soil, and lastly, the degradative plasmids can get lost in the soil.

Enzymatic bioremediation is another advantageous method used for cleaning up polluted soils. Degradation of PAHs can be achieved mainly through oxygenase, dehydrogenase, phosphatases, and ligninolytic enzymes. It has been reported that most degradative enzymes work optimally at mesophilic temperatures and their activity decreases at very high and low temperatures. Some enzymes, such as laccase and Mn-dependent peroxidase, are exceptions to this and can be active at temperatures as low as 5°C and as high as 75°C, respectively [44]. Unlike the general notion of substrate specificity, ligninolytic enzymes act non-specifically on phenolic and non-phenolic organic compounds *via* the generation of cation radicals after one-electron oxidation [45].

The final technique is rhizoremediation, which uses both plants and their associated rhizospheric microorganisms. It was found effective in accelerating the degradation of organic pollutants as (i) plants provide a large surface area as well as facilitate the microbes reaching deep into the soil; (ii) they allow selective enrichment of microorganisms for the degradation of xenobiotics in root-free soils [46]; (iii) roots induce co-metabolism by supplying nutrients such as amino acids, carbohydrates, and organic acids [47]; and (iv) they provide co-factors essential for degradative enzyme activation [4, 48].

Table 1. Applications of soil microorganisms in bioremediation of xenobiotic compounds.

S. No.	Organism	Applications	References
BACTERIA			
1.	*Pseudomonas* spp.[A] *P. fluorescens*[B] and *P. putida*[C]	Bioremediation of PAH's (dioxygenase)[A], heavy metals[C] (Cr (VI),Hg), naphthalene[C], pesticides[C] (2,4-D), and rhizoremediation of PCBs[B] (with sugar beet)	[38, 49 - 54]
2.	*Burkholderia cepacia*	Rhizoremediation of 2,4,-D with barley	[55]
3.	Actinomycetes	Rhizoremediation of 1,4-dioxane with *Populus deltoides nigra*	[56]

(Table 1) cont.....

S. No.	Organism	Applications	References
4.	*Stenotrophomonas maltophilia*	Bioremediation of high molecular weight PAHs, such as methomyl (pesticide)	[57]
6.	*Xanthobacter autotrophicus*	Enzymatic bioremediation of PAH's (dehaloginase)	[58]
7.	*Mycobacterium* spp.	Enzymatic bioremediation of PCB's (Laccase)	[59]
8.	*Comamonas* spp.	Enzymatic bioremediation of nitroaromatic compounds (*e.g.*, 2,4,6-trinitotoluene) (nitroreductase)	[60]
9.	*Kurthia* sp. *Micrococcus* sp. *Bacillus* sp. *Dienococcus* sp. and endophytic *Bacillus* sp.	Rhizoremediation of PAH's with *Populus deltoides*	[61, 62]
11.	*Sporosarcina ginsengisoli*	Arsenic [As (III)]	[63]
12.	*Bacillus cereus* strain XMCr-6 and *B. subtilis*	Cr (VI)	[54, 64]
14.	*Enterobacter cloacae* B2-DHA	Cr (VI)	[65]
15.	Consortia of *Viridibacillus arenosi* B-21, *Sporosarcina soli* B-22, *Enterobacter cloacae* KJ-46, and *E. cloacae* KJ-47	Pb, Cd, and Cu	[66]
16.	*Deinococcus geothemalis*	Hg at high temperature	[67]
17.	*Cupriavidus metallidurans* strain MSR33	Hg biodegradation along with the synthesis of organomercurial lyase protein (merB) and mercuric reductase (merA)	[68]
18.	*Bacillus* sp., most aerobic bacteria, and all fungi	Enzymatic bioremediation of aromatic and aliphatic hydrocarbon (cytochrome P450 mono-oxygenase)	[69]
FUNGI			
19.	*Phanerochaete chrysosporium*	Enzymatic bioremediation of aliphatic hydrocarbons (peroxidase)	[70]
20.	*Aspergillus niger* and *A. fumigatus*	Enzymatic bioremediation of aromatic and aliphatic nitriles (nitrilase) Pb	[71, 72]
21.	*Gloeophyllum sepiarium*	Cr (VI)	[73]
22.	*Rhizopus oryzae* (MPRO)	Cr (VI)	[74]

Abbreviations **PCBs:** Polychlorinated biphenyls; **PAHs:** Polycyclic aromatic hydrocarbons; **2,4-D:** 2,4-Dichlorophenoxyacetic acid.

Future Aspects

Further research should not only focus on identifying new species with bioremedial potential but should also aim at developing strategies to improve the degradative capabilities of microorganisms that are already being used in bioremediation. Actinomycetes have been reported to degrade pesticides with widely different chemical structures, such as organochlorines, striazines, triazinones, carbamates, organophosphates, organophosphonates, acetanilides, and sulfonylureas. However, very little is known about the mechanisms governing these biotransformations. Many fungi, like *Phanerochaete* spp., can be potentially utilized to degrade pollutants that cannot be achieved through bacterial remediation. However, techniques must be explored to hasten such transformations [5]. Genetically modified organisms have been shown to efficiently degrade many different types of pollutants, though the field-testing of these organisms is subject to approval by regulatory committees [75].

SOIL MICROBES IN BIODETERIORATION

As described in the previous section, degradation is vital to the phenomena of recycling and occurs naturally. It can be the weathering of rocks, degradation of organic materials like wood and paper, solubilization of minerals, or the slow but continuous degradation of plastic waste. However, unlike biodegradation, crucial activities of organisms in biofilms also play a role in the undesirable alteration of a range of economically and/or culturally important materials in a process known as biodeterioration. This list includes metals, textiles, paper and pulp, wood, leather, stone, polymers, concrete, food, *etc.*, and materials made thereof. The susceptibility of a substratum to this negative facet of microbial activity in soil depends on various surface and environmental factors, including the type of material, humidity, and the indigenous and foreign microflora [76, 77]. Sand [78] enlisted nine broad means by which microbes can lead to biodeterioration: (i) physical presence; (ii) forming biofilms; (iii) secreting exoenzymes; (iv) forming noxious compounds; (v) creating salt stress; (vi) producing organic acids; (vii) producing inorganic acids; (viii) generating chelating agents; and (ix) secreting organic solvents. We will elaborate on this in the following section.

Biodeterioration

Biodeterioration is the negative aspect of the biodegradative ability of microorganisms, leading to huge economic losses. It is defined as any undesirable, irreversible change in the properties of any useful and valuable material due to the metabolic activity of microorganisms called biodeteriogens. Every biodegradable material is prone to biodeterioration right from its raw materials to the finished product stage and commonly includes food, textiles, paper, leather, and wood. Materials like metal, stone, plastics, *etc.*, which are not directly degraded by microorganisms, are subjected to microbial deterioration indirectly through the formation of microbial biofilms on them and the secretion of corrosive products by microorganisms.

Woods

The main structural components of wood are cellulose and hemicellulose (collectively known as holocellulose), and lignin and wood are very low in nitrogen content. These factors, along with the presence of tannins (in broad-leaved trees) and other phenolic compounds (in conifers), make it quite a tough substrate to undergo degradation. Despite these barriers, wood is still vulnerable to attack by a variety of agents, the most common of which are fungi. They can colonize and decay wood tissues in trees, cut wooden logs (timber) to process wood for sale (lumber), and finish indoor and outdoor wood products such as furniture and poles, respectively. Wood decay fungi result in a loss of their strength and are categorized as soft-rot, brown-rot, and white-rot fungi (Table **2**). Even though colonization by wood staining fungi and other moulds increases moisture uptake and induces slight strength loss, they are generally not considered decay of wood [79].

Table 2. Wood decaying fungi [79, 81].

S. No.	Fungi	Example	Advanced Degradation Stage	Enzymes Secreted	Comments
1.	**Soft-rot fungi**	*Chaetomium* and *Ceratocystis*	Softer than intact wood; Brown, weathered appearance; and Brittle surface	Cellulosic enzymes (endo-1,4-glucanase, exo-1,4-β-glucanase, and 1,4-β-glucosidase)	1. Attack wood in a moist environment. 2. Access lumber after invading pits on the wall and passing fine microhyphae through a thin S3 lignin layer. On reaching the cellulose-rich S2 layer, fungi produce broad, branched 'T-shaped' hyphae in longitudinal planes. 3. The enzymes and degradative metabolites are then secreted and result in characteristic rhomboidal cavities depicting cellulose degradation. 4. The lignin layer remains largely unaffected.

(Table 2) cont.....

S. No.	Fungi	Example	Advanced Degradation Stage	Enzymes Secreted	Comments
2.	**Brown-rot fungi**	*Poria placenta*, *Coniophora puteana*, and *Serpula lacrymans* (dry rot)	Brown, crumbly, with brick-like cracks	1. Endoglucanases: cleaves β-1,4-glycosidic bond. 2. β-glucosidases: hydrolyse cellobiose 3. Endo-xylanases and β-xylosidase: breakdown hemicellulose 4. *C. puteana* can also produce cellobiohydrolase	1. Fungi degrade the holocellulose component and leave lignin unutilized; hence, the substrate appears brown. 2. Brown-rot fungi can bore holes through cell walls and secrete both enzymatic and non-enzymatic metabolites. 3. During initial decay stages, extensive cellulose depolymerization causes early loss in strength. 4. Enzymes are too large to pass through small pores and causes an extensive depolymerization in the initial fungal attacks. Fungi produce low molecular weight quinones, catechols, and hydroxybenzene derivatives to induce non-enzymatic degradation of cell walls. *Gloeophyllum trabeum* produces phenolate compounds to serve this purpose.

(Table 2) cont.....

S. No.	Fungi	Example	Advanced Degradation Stage	Enzymes Secreted	Comments
3.	**White-rot fungi**	*Phellinus pini* and *Ganoderma applanatum*	Bleached appearance	1. Lignolytic: LiP, laccase, and MnP. 2. Endoglucanases, β-glucosidases, and cellobiohydrolase: degrade holocellulose 3. Xylosidase, xylanase, *etc.*, (for a complete breakdown of hemicellulose)	1. Generally, white-rot fungi degrade hemicellulose, cellulose, and lignin at a uniform rate in 'simultaneous white rot' of wood. 2. However, sometimes hemicellulose and lignin can be preferentially degraded to cause selective white rot of wood. 3. Fungi oxidize lignin by producing ligninolytic enzymes. LiP can oxidize both phenolic and non-phenolic lignin but cannot penetrate intact cell walls of wood due to its size. MnP oxidizes Mn^{+2} to Mn^{+3}, which further oxidizes organic molecules. Whereas laccase oxidizes lignin while simultaneously reducing oxygen to produce water. 4. Non-enzymatic mechanisms are less explored. However, their activities have been observed, particularly in selective white rot.

Abbreviations: LiP: Lignin peroxidase; MnP: Manganese peroxidase; S2 and S3: Secondary cell wall layers.

While brown-rot and white-rot fungi belong to Basidiomycetes and are described by the appearance of the substrate upon advanced deterioration, soft-rot fungi are typically Ascomycetes or fungi-imperfecti (Deuteromycetes) and are named so to describe the soft surface of the wood on degradation. Moreover, wood-staining fungi such as *Aureobasidium*, *Alternaria*, and *Cladosporium* often settle in the rays of wood and cause bluish or brownish discoloration. Although they do not cause decay, the staining of timber significantly affects its value. Optimum temperature, pH, and moisture content promote these deterioration processes. However, during unfavourable conditions, fungi can still form spores or other resistant structures and remain viable. Furthermore, wood is also prone to deterioration by cellulolytic bacteria like *Cytophaga*, *Cellvibrio* (aerobic), and *Clostridium* (anaerobic) [80].

Papers

Papers have high cellulose content other than lignin, hemicellulose, pectins, proteins, *etc.*, and provide favourable nutritional conditions for microbial attack. Bacteria, fungi, and actinomycetes thrive on paper, wreaking havoc in libraries and archives [80]. Fungi are the most frequent among all and even pose a health risk to professionals. A high number of common outdoor fungal flora (*Penicillium*, *Aspergillus*, *etc.*) were reported in the air of archives and libraries. However, the frequency of *Fusarium* spp., *Chaetomium* spp., and *Geotrichum* spp. was unusually higher than in the outside air. Many of the fungi reported frequently from libraries are capable of degrading cellulose (*Chaetomium* spp. and *Trichoderma* spp.) and hence could weaken the structure and integrity of papers. Moreover, fungi promote hydrolysis of paper fibres by secreting acids and other extracellular enzymes, primarily cellulose as cellulose is the main component in papers, but also amylase and protease as starch and proteins are added as binders and in paper coatings, *etc.* [82]. Some can even produce stains of different colours (red, yellow, brown, *etc.*) and varied shapes (round or irregular) and/or secrete tannase, which can decolourize ink (*Aspergillus* and *Penicillium*) [80].

Leather and Textile

Leather manufacturing is prone to attack by biodeteriogens (due to the ample presence of proteins and lipids) at different stages, right from raw hides to storage of finished products. Bacteria generally participate before the product is tanned: in rawhides (*Micrococcus* forms coloured spots) and during soaking (*Bacillus subtilis*, *Pseudomonas aeruginosa*, *etc.*), while fungi (*Penicillium*, *Aspergillus*, *Mucor*, *Rhizopus*, *etc.*) can degrade even tanned leather. Some tannery agents contain compounds that can hinder the growth of bacteria [83].

Textiles are also susceptible to varying degrees of deterioration. The ones made of natural fibres are relatively more susceptible to microbial attack, while synthetic ones are rather resistant. Depending on humidity, temperature, the reactivity of material, *etc.*, fabrics are attacked by different biodeteriogens. However, fungi are the most frequent culprits. Fungi such as *Chaetomium*, *Trichoderma*, *Penicillium*, *Aspergillus*, *etc.*, can degrade cellulose fabrics, while *Microsporum*, *Fusarium*, *etc.*, can degrade keratin in wool fabrics. Although less significant, bacteria (*Bacillus*, *Pseudomonas*, *Clostridium*, *etc.*) can also degrade textiles [84].

Iron

Due to its abundance in ores and reliable mechanical properties, iron is widely used in various sectors. A major downside, however, to this widespread usage is its vulnerability to corrosion, subsequently causing huge economic losses [77].

Microorganisms (both aerobic and anaerobic) (Table **3**) can cause microbiologically induced corrosion (MIC), resulting in the release of metals in solution. Under aerobic conditions, *Thiobacillus*, *Gallionella*, and *Leptothrix* oxidize iron in an energy-yielding step to form tubercles (corrosion products) with three distinct layers, *viz.*, (a) the outer orange layer has ferric hydroxide (Fe[OH]$_3$), (b) the innermost green layer has ferrous hydroxide (Fe[OH]$_2$), and (c) the middle layer contains black magnetite (Fe$_3$O$_4$) [76, 85]. However, MIC is dramatically accelerated and is more pronounced when conditions are nearly anoxic with pH 6. Under these conditions, sulfate-reducing bacteria (SRB) reduce sulfate to hydrogen sulfide by employing electrons from molecular hydrogen or degradation of organic matter to gain energy. The noxious compounds like H$_2$S and other corrosive sulfides then rapidly react with metallic iron and form a characteristic iron sulfide (chemical microbially influenced corrosion, CMIC).

Recently, it was found that iron can also function as a sole electron donor (while being oxidized to ferrous ions) to reduce sulfur at a high rate. Electrical microbially influenced corrosion (EMIC) is exhibited by a limited number of SRB, including *Desulfopila corrodens* and *Desulfovibrio ferrophilus*, while all other SRB (*e.g.*, *Desulfobacterium catecholicum*, *Desulfobulbus elongatus*, *D. japonicus*, *Desulfovibrio desulfuricans*, *etc.*) demonstrate CMIC [77]. *Archaeoglobus fulgidus*, an archaeal thermophile, and *Methanosarcina barkeri*, methanogenic bacteria, are also shown to promote corrosion [85]. Furthermore, *Shewanella* spp. and *Geobacter* spp. can corrode carbon steel under anoxic conditions [86].

Table 3. Microorganisms involved in the corrosion of iron [76, 77, 85, 87].

Aerobic		Anaerobic
Acholeplasma	*Leptothrix*	*Desulfobacter*
Acinetobacter	*Naumanniella*	*Desulfobulbus*
Actinomyces	*Pedomicrobium*	*Desulfococcus*
Archangium	*Seliberia*	*Desulfomonas*
Arthrobacter	*Siderocapsa*	*Desulfonema*
Crenothrix	*Siderococcus*	*Desulfosarcina*
Ferrobacillus	*Sphaerotilus*	*Desulfotomaculum*
Gallionella	*Thiobacillus*	*Desulfovibrio*
Herpetosiphon		*Methanosarcina barkeri*

Although anaerobic sulfate-reducing bacteria are predominant in anoxic aquatic environments like marine sediments, swamps, and lake depths, they can also be widely present in anoxic soil and some are even found in the intestines of humans and animals.

Stone

Weathering of rocks is an important event in the formation of soil and routinely occurs due to various abiotic factors like wind, rain, salts, temperature fluctuations, *etc.* The association of microbial activity with the weathering process accelerates the same. Unfortunately, this also leads to the deterioration of monuments, tombs, and other cultural heritage, including rock paintings that are already badly preserved. Microbial communities on rocks can contribute to both biophysical and biochemical deterioration. The former results from penetration of pre-existing pores, creating stress that leads to physical damage like fragmentation, while the latter causes dissolution of rock minerals due to the action of inorganic and organic acids. The inorganic nature of rocks does not support the growth of heterotrophic communities like bacteria and fungi. Therefore, photoautotrophic organisms like lichens (*Roccella*, *Candelaria*, *Leptotrema*, *Parmelia*, *etc.*) play a pioneering role in the colonization of rocks by fixing inorganic carbon and facilitating the further succession of more corrosive biodeteriogens. Receptivity (ability to be colonized) of stones largely depends on moisture content, chemical composition, texture, and the existing microclimate [88]. Chemolithoautotrophic bacteria like sulfur-oxidizing *Thiobacillus* produce sulfates that get inside the pores with rainwater and exert pressure on recrystallization, resulting in salt stress. Nitrifying bacteria like *Nitrosomonas*, *Nitrosococcus*, and *Nitrobacter* oxidize ammonia to nitrites, which further form nitric acid. The production of such inorganic acid(s) can result in stone dissolution and powdering. Chemoorganotrophic bacteria, on the other hand, can produce several organic acids that can chelate metal cations and cause stone dissolution. Cyanobacteria like *Anabaena*, *Aulosira*, *Entophysalis*, *Gloeocapsa*, *Nostoc*, and *Scytonema* produce coloured biofilms on buildings, causing major aesthetic damage. They enhance water-holding capacity and retain dust, spores, and other contaminants on the surface. Additionally, bacteria belonging to the genera *Arthrobacter*, *Bacillus*, *Desulfovibrio*, *Pseudomonas*, *Micrococcus*, and *Staphylococcus* have also been reported from decaying stones and rock paintings [80, 88, 89].

Further, fungi can heavily colonize rocks and are believed to be the most detrimental. Several genera of fungi have been reported from monuments and rock paintings, including *Aspergillus*, *Acremonium*, *Alternaria*, *Aureobasidium*, *Beauveria*, *Botrytis*, *Chaetomella*, *Chaetomium*, *Chrysosporium*, *Cladosporium*,

Cunninghamella, Curvularia, Emericella, Epicoccum, Fusarium, Lasiodiplodia, Penicillium, Phoma, Rhizopus, Trichoderma, and *Verticillium* [80, 89, 90]. Fungal species can corrode stones chemically and mechanically by producing organic acids (chelate metal cations) and protruding their hyphae and spores onto the rock surface. They can also stain stones by oxidizing manganese, causing aesthetic decay. Furthermore, microbes can produce organic solvents like ethanol and ketone, which can partially dissolve several substances [78].

Polyethylene and Plasticized PVC

The very omnipresence of microorganisms enables them to inhabit even the most unlivable substratum. This susceptibility of the substratum plays a crucial role in alleviating some of the major environmental woes, although sometimes very slowly. For instance, the tonnes of plastic waste that we generate every day. This, however, jeopardises their use in other structures, such as pipes and storage tanks [91]. Microorganisms colonize surfaces in mixed cultures and reside in them by forming biofilms, which give them added stress resistance. Webb and coworkers [92] discovered that *Aureobasidium pullulans* secrete extracellular polysaccharides in order to adhere to and use plasticized polyvinyl chloride (pPVC) as their sole carbon source. Other than producing extracellular esterase to degrade dioctyl adipate (DOA) in pPVC, *A. pullulans* can also use exogenous carbon sources that settle on the substratum over time by secreting exoenzymes like cellulase, invertase, phosphatase, proteinase, and maltase. Furthermore, filamentous fungi from the genera *Alternaria, Aspergillus, Penicillium, Cladosporium, Emericella, Paecilomyces* and yeasts *Rhodotorula,* and *Kluyveromyces* have been observed colonising pPVC in succession [92]. Sometimes these exoenzymes are required to degrade high molecular weight polymers such that they can then be taken up and metabolized by the organism [91].

Control of Biodeterioration

The invading biodeteriogens, the surface of interest, and the existing environmental conditions account for three main aspects of successful biodeterioration. Many of these aspects should be able to be changed with proper prevention and control techniques [80]. An ideal product in this regard should have broad-spectrum activity, low or no toxicity, low cost, and should be preferably biodegradable. Moreover, they should not react with the surface and contribute to its further corrosion.

Wood

Optimum moisture content is one of the primary determinants of fungal attacks on wood. Using seasoned wood and ensuring the non-availability of enough moisture around both the timber and finished products could save a great sum of money in the maintenance stages. Certain chemical alternatives have also been developed to prevent wood decay. Preservatives such as creosote, zinc chloride, and others are commonly used in pressure treatment. However, some, such as pentachlorophenol, are no longer recommended due to their high toxicity. Certain fungicidal agents are also available to control the growth on existing surfaces. Boron-containing products (borates) are less hazardous, long-lasting, water-soluble agents that are potent in controlling both wood-destroying fungi and insects. A more environmentally friendly approach to this menace is now offered by *Trichoderma* spp. This fungus grows relatively faster and competes with decay fungi for nutrients. They can also produce antifungal compounds to limit the growth of wood-decaying fungi. Several bacteria and actinomycetes are also being investigated to serve similar purposes [81]. Coating of finished products with paints or polish, which may or may not have antimicrobials incorporated in them, significantly reduces the deterioration.

Paper and Pulp

Several physical and chemical methods are used to control and prevent the deterioration of paper. Physical methods like dehydration, irradiation with UV and gamma-rays, or using extreme temperatures can exert an immediate effect, but they are usually short-lived. Whereas chemical methods leave residues after the action, which extends their antimicrobial effect. Alcohols, phenols, quaternary ammonium salts, and azole derivatives are commonly used as microbicides. These membrane-active microbicides get absorbed on the cell wall surface and alter it to allow access to internal components of cells to exhibit their lethal effect. While alcohols are used as disinfectants, alkylating agents like ethylene oxide and formaldehyde are used for fumigating archives and museums and have sporicidal activity as well. These fumigants, however, are also toxic to human health. Quaternary ammonium compounds (quats) are cationic and act to alter cell wall permeability by forming electrostatic bonds with negatively charged sites on the cell wall. This leads to the entry of active ingredients inside the cell and results in cell death. Quats inhibit spore germination and are hence sporostatic. Several natural plant extracts are now also used for the conservation of heritage objects [93].

Another cause of concern for the pulp and paper industries is the slime deposited by microbes on machinery. Due to structural heterogeneity, these structures show increased resistance to biocides and cause deterioration of both raw materials and finished products. The use of slimecides along with mechanical cleaning and boil-outs at extreme temperatures and extreme pH has hence become an indispensable part of paper production [94].

Leather and Textiles

To control the deterioration of leather, leather industries previously used fungicides such as pentachlorophenol, which has now been replaced due to toxicity and poor degradability with biocides such as 2-(thiocyanomethylthio) benzothiazole (TCMTB) [83].

Moreover, controlling physical conditions during storage can significantly prevent the deterioration of finished products and extend their lives. This also holds true for the textile industry. However, biocides may also be used while manufacturing fabrics that are predicted to be used in conditions extremely favorable for the growth of microbes (tropical region) [84].

Metal

Biocides are conventionally used to get rid of biofilms on metal surfaces and can be both oxidizing and non-oxidizing. Oxidizing biocides like chlorine are frequently used in pipes associated with water utilities owing to their low cost and excellent efficacy. While non-oxidizing biocides like glutaraldehyde, quats and triazines offer broad-spectrum activity and are used in the oil and gas industries. Moreover, potassium dichromate, organoborane, *etc.* are used to prevent fungal corrosion in jet fuel tanks. Further, physical control methods are also employed to instil a barrier over metal surfaces. These include methods like applying the protective coating and cathodic protection. In the latter, a sacrificial metal is used, which itself undergoes corrosion and protects the metal (of interest) by supplying electrons [95].

Stone

The attack of biodeteriogens requires optimum environmental conditions. Although these are beyond human control in a natural environment, they can, to some extent, be controlled indoors. Further, periodic cleaning of the surface could remove nutritive factors (bird droppings, pollutants, *etc.*) that microbes thrive on. Consolidation of weakened stones on monuments and the application of protective paints on modern buildings can prevent and delay further deterioration. However, once established, biofilms will be required to be removed by the application of biocides, as already discussed in previous sections [80].

Future Aspect

Monuments, historic literature, paintings, and artefacts serve as permanent records of early human civilization, and their need for conservation is indispensable. Knowledge of biodeteriogens, their mechanisms of action, and suitable environmental conditions are of immense importance to the development of strategic conservation approaches. Similarly, modern structures using metals, polymers, and concrete are also prone to deterioration. Rapid industrialization and urbanization have polluted the environment in numerous ways and have further worsened the problem. There are three main aspects of biodeterioration: the invading microbes, the surface of interest, and the existing environmental conditions. Investigating and coming up with the methods to alter these aspects can influence the growth of biodeteriogens and hence mitigate the loss of valuable materials. Moreover, a greater understanding of degradation processes has also found application for some commercial purposes. For example, white-rot fungi like *Phlebia subseries* and *Ceriporiopsis subvermispora* are used as bio-pulping agents and turn down energy as well as environmentally noxious chemical requirements for mechanically refining wood chips. Moreover, *Phanerochaete chrysosporium* and *P. crassa* are being investigated as bio-bleaching applicants.

CONCLUDING REMARKS

Soil microorganisms mediate nutrient transformations that result in enhanced soil fertility and better nutrient acquisition by plants. We saw how biological nitrogen fixers like *Rhizobium* and *Azotobacter* employ nitrogenase enzyme complexes to fix elemental nitrogen into an organic form, which is then assimilated and converted into amino acids and proteins. A diverse array of bacteria can even convert these organic forms of nitrogen to ammonia, and yet others can further oxidize ammonia into nitrite and nitrate. Following this, denitrifiers like *Pseudomonas* reduce these nitrates back to elemental nitrogen and replenish atmospheric sources. Similarly, phosphorus in soils is present as insoluble organic components in humus or living cells, or as insoluble inorganic salts. Phosphate solubilizing microorganisms (*e.g.*, *Pseudomonas*) employ distinct mechanisms like producing organic and inorganic acids and/or enzymes like phytases, phosphatases, and phosphonatases to solubilize and mineralize these unavailable forms of phosphorus. However, to feed a growing population, we still rely on chemical fertilizers to supplement the plant's nutritional requirements. This bestows a wide prospect of identifying functional strains and elucidating model systems that can be scaled up as an alternative to agrochemicals.

Furthermore, the biodegradation of these and many other environmental contaminants has been an active concern lately. Industrialization has led to the accumulation of toxic xenobiotics, mainly hydrocarbons, chlorinated solvents, PCBs, PAHs, and heavy metals. Bioremediation is an eco-friendly, cost-effective solution that has been in use for the past few decades and has been shown to reduce the number of contaminants to acceptable, non-toxic, or undetectable levels, and in some cases, even achieve mineralization of the pollutants. However, carrying out bioremediation in a limited time frame and maintaining high numbers of degrading microorganisms has proven to be a challenge. Therefore, establishing a stable microbial ecosystem is vital in implementing bioremediation engineering projects and requires further research.

These degradative potentials of microorganisms are sometimes the only hope to mitigate several major environmental pollutants, like the ever-growing menace of plastic waste. However, these polymers, owing to their unique characteristics, are also widely used in other important structures and their susceptibility to microbial action is a cause of concern here. In the final section of this chapter, we shifted our focus to this detrimental aspect of soil microbes. Their presence in biofilms gives them added tolerance to environmental stress, and this, coupled with a wide array of secreted enzymes, allows them to inhabit several substrata. Though the heterotrophic communities are incapable of colonizing stones, the pioneer phototrophs turn the surface favourable for succession by fixing inorganic carbon. Cyanobacteria on buildings also enhance the water holding capacity and retain contaminants, causing aesthetic damage. Fungi, on the other hand, synthesize organic acids and can also protrude their hyphae into rock pores. Moreover, microbiologically induced corrosion (MIC) causes the dissolution of metals and can occur in both aerobic and anaerobic conditions. In the former, *Thiobacillus* oxidizes iron and forms tubercles in an energy-yielding process, whereas SRB corrodes metal surfaces under anoxic conditions by producing noxious compounds. Thus, the process of biodeterioration causes huge losses in terms of repairing and maintaining these structures and mandates large-scale investigation to foster ways to alleviate them and further diminish knowledge gaps.

CONSENT FOR PUBLICATION

Not applicable.

CONFLICT OF INTEREST

The author declares no conflict of interest, financial or otherwise.

ACKNOWLEDGEMENTS

The authors would like to thank Dr. Rakesh Kumar Gupta, Principal, Ram Lal Anand College, for providing all the necessary infrastructure and support in compiling this chapter.

REFERENCES

[1]   Galloway JN, Leach AM, Bleeker A, *et al.* A chronology of human understanding of the nitrogen cycle. Philos Trans R Soc Lond B Biol Sci 2013; 368: 20130120.
        [http://dx.doi.org/10.1098/rstb.2013.0120] [PMID: 23713118]

[2]   Doran JW, Safley M. Defining and Assessing Soil Health and Sustainable Productivity. Pankhurst CE, Doube BM, Gupta VVSR, Eds. Biological Indicators of Soil Health. Wallingford: CAB International 1997; pp. 1-28.

[3]   Bisen PS, Debnath M, Prasad GBKS, Eds. Microbes: Concepts and Applications. Blackwell John Wiley and Sons 2012; p. 724.

[4]   Alexander M II, Ed. Biodegradation and Bioremediation. San Diego, CA: Academic Press 1999; p. 453.

[5]   Nawaz K, Hussain K, Choudary N, *et al.* Eco-friendly role of biodegradation against agricultural pesticides hazards. Afr J Microbiol Res 2011; 5: 177-83.

[6]   Rahman KS, Rahman T, Lakshmanaperumalsamy P, *et al.* Occurrence of crude oil degrading bacteria in gasoline and diesel station soils. J Basic Microbiol 2002; 42: 284-91.
        [PMID: 12210553]

[7]   Saxena S, Ed. Applied Microbiology. Springer 2015; p. 190.

[8]   Wiley JM, Sherwood LM, Woolverton CJ. Prescott's Microbiology. 9th ed. McGraw Hill International 2013; p. 1152.

[9]   Pepper IL, Gerba CP, Gentry TJ, III, Eds. Environmental Microbiology. UK: Elsevier 2015; p. 728.

[10]   Robertson GP, Groffman PM. Nitrogen Transformation. Paul EA, Ed. Soil Microbiology, Biochemistry and Ecology. New York, USA: Springer 2007; pp. 341-64.

[11]   Wagner SC. Biological nitrogen fixation. Nature Education Knowledge 2012; 3: 15.

[12]   Vitousek PM, Cassman K, Cleveland C, *et al.* Towards an ecological understanding of biological nitrogen fixation. Biogeochemistry 2002; 57: 1-45.

[13]   Atlas RM, Bartha R, IV, Eds. Microbial Ecology: Fundamentals and Applications. USA: Benjamin Cumming Science Publishing 2000; p. 576.

[14]   Gottschalk G, Ed. Bacterial Metabolism. New York: Springer-Verlag 1979; p. 359.

[15]   Payne WJ. Reduction of nitrogenous oxides by microorganisms. Bacteriol Rev 1973; 37: 409-52.
        [PMID: 4203393]

[16]   Humbert S, Tarnawski S, Fromin N, *et al.* Molecular detection of anammox bacteria in terrestrial ecosystems: distribution and diversity. ISME J 2010; 4: 450-4.
        [PMID: 20010634]

[17]   Huang L, Jia X, Zhang G, *et al.* Soil organic phosphorus transformation during ecosystem development: a review. Plant Soil 2017; 417: 17-42.

[18]   Yu X, Doroghazi JR, Janga SC, *et al.* Diversity and abundance of phosphonate biosynthetic genes in nature. Proc Natl Acad Sci USA 2013; 110: 20759-64.
        [PMID: 24297932]

[19]   Bhattacharya A, Ed. Changing Climate and Resource Use Efficiency in Plants. UK: Academic Press 2019; p. 324.

[20]   Alori ET, Glick BR, Babalola OO. Microbial phosphorus solubilization and its potential for use in sustainable agriculture. Front Microbiol 2017; 8: 971.
[http://dx.doi.org/10.3389/fmicb.2017.00971] [PMID: 28626450]

[21]   Kalayu G. Phosphate solubilizing microorganisms: promising approach as biofertilizers. Int J Agron 2019; 2019: 4917256.
[http://dx.doi.org/10.1155/2019/4917256]

[22]   Sharma SB, Sayyed RZ, Trivedi MH, *et al.* Phosphate solubilizing microbes: sustainable approach for managing phosphorus deficiency in agricultural soils. Springerplus 2013; 2: 587.
[http://dx.doi.org/10.1186/2193-1801-2-587] [PMID: 25674415]

[23]   Eida MF, Nagaoka T, Wasaki J, *et al.* Phytate degradation by fungi and bacteria that inhabit sawdust and coffee residue composts. Microbes Environ 2013; 28: 71-80.
[PMID: 23100024]

[24]   Jain S, Arnepalli D. Biomineralization as a Remediation Technique: A Critical Review. Stalin V, Muttharam M, Eds. Geotechnical Characterisation and Geoenvironmental Engineering Lecture Notes in Civil Engineering. Singapore: Springer 2016; Vol. 16: pp. 155-62.

[25]   Mervat SM. Degradation of methomyl by the novel bacterial strain *Stenotrophomonas maltophilia* M1. Electron J Biotechnol 2009; 12: 1-6.

[26]   Singh DK. Biodegradation and bioremediation of pesticide in soil: concept, method and recent developments. Indian J Microbiol 2008; 48: 35-40.
[PMID: 23100698]

[27]   Chung MJ, Jong KA. Isolation and characterization of 2, 4- dichlorophenoxyacetic acid degrading bacteria from paddy soils. J Microbiol 1998; 36: 256-61.

[28]   Laemmli CM, Leveau JHJ, Zehnder AJB, *et al.* Characterization of a second tfd gene cluster for chlorophenol and chlorocatechol metabolism on plasmid pJP4 in *Ralstonia eutropha* JMP134(pJP4). J Bacteriol 2000; 182: 4165-72.
[PMID: 10894723]

[29]   Sayler GS, Hooper SW, Layton AC, *et al.* Catabolic plasmids of environmental and ecological significance. Microb Ecol 1990; 19: 1-20.
[PMID: 24196251]

[30]   Aislabie J, Lloyd-Jones G. A review of bacterial-degradation of pesticides. Aust J Soil Res 1995; 33: 925-42.

[31]   Mulchandani A, Kaneva I, Chen W. Detoxification of organophosphate pesticides by immobilized *Escherichia coli* expressing organophosphorus hydrolase on cell surface. Biotechnol Bioeng 1999; 63: 216-23.
[PMID: 10099598]

[32]   Richins RD, Kaneva I, Mulchandani A, *et al.* Biodegradation of organophosphorus pesticides by surface-expressed organophosphorus hydrolase. Nat Biotechnol 1997; 15: 984-7.
[PMID: 9335050]

[33]   Lodewyckx C, Vangronsveld J, Porteus F, *et al.* Endophytic bacteria and their potential application. Crit Rev Plant Sci 2002; 6: 583-606.

[34]   Stone JK, Bacon CW, White JF. An Overview of Endophytic Microbes: Endophytism Defined. Bacon CW, White JF, Eds. Microbial Endophytes. New York: Marcel Dekker 2000; pp. 3-29.

[35]   Sessitsch A, Reiter B, Pfeifer U, *et al.* Cultivation-independent population analysis of bacterial endophytes in three potato varieties based on eubacterial and Actinomycetes-specific PCR of 16S rRNA genes. FEMS Microbiol Ecol 2002; 39: 23-32.

[PMID: 19709181]

[36]   Cameron MD, Timofeevski S, Aust SD. Enzymology of *Phanerochaete chrysosporium* with respect to the degradation of recalcitrant compounds and xenobiotics. Appl Microbiol Biotechnol 2000; 54: 751-8.
[PMID: 11152065]

[37]   Premjet S, Bunthong O, Tachibana S, *et al.* Screening of fungi from natural sources in Thailand for degradation of polychlorinated hydrocarbons. Am-Eurasian J Agric Environ Sci 2009; 5: 466-72.

[38]   Kuiper I, Lagendijk EL, Bloemberg GV, *et al.* Rhizoremediation: a beneficial plant-microbe interaction. Mol Plant Microbe Interact 2004; 17: 6-15.
[PMID: 14714863]

[39]   Widada J, Nojiri H, Omori T. Recent developments in molecular techniques for identification and monitoring of xenobiotic-degrading bacteria and their catabolic genes in bioremediation. Appl Microbiol Biotechnol 2002; 60: 45-59.
[PMID: 12382041]

[40]   Ramadan MA, el-Tayeb OM, Alexander M. Inoculum size as a factor limiting success of inoculation for biodegradation. Appl Environ Microbiol 1990; 56: 1392-6.
[PMID: 2339892]

[41]   Prescott LM, Harley JP, Klein DA. 5th ed. Microbiology. The McGraw-Hill Companies, Inc, North America 2002; 1026.

[42]   Swaminathan K, Subrahmanyam PV. Biodegradation of p-nitrophenol in anaerobic fixed film fixed bed reactor. Indian J Environ Health 2002; 44: 8-11.
[PMID: 12968719]

[43]   Gentry TJ, Rensing C, Pepper LI. New approaches for bioaugmentation as a remediation technology. Crit Rev Environ Sci Technol 2004; 34: 447-94.

[44]   Farnet AM, Criquet S, Tagger S, *et al.* Purification, partial characterization, and reactivity with aromatic compounds of two laccases from *Marasmius quercophilus* strain 17. Can J Microbiol 2000; 46: 189-94.
[PMID: 10749532]

[45]   Wilson SC, Jones KC. Bioremediation of soil contaminated with polynuclear aromatic hydrocarbons (PAHs): a review. Environ Pollut 1993; 81: 229-49.
[PMID: 15091809]

[46]   Nichols TD, Wolf DC, Roders HB, *et al.* Rhizosphere microbial populations in contaminated soils. Water Air Soil Pollut 1997; 95: 165-78.

[47]   Gupta A, Joia J, Sood A, *et al.* Lindane and its degradation from environment using biotechnological approach: a review. Int J Recent Sci Res 2016; 7: 13756-65.

[48]   Haby PA, Crowley DE. Biodegradation of 3-chlorobenzoate as affected by rhizodeposition and selected carbon substrates. J Environ Qual 1996; 25: 304-10.

[49]   Tsuda M, Iino T. Naphthalene degrading genes on plasmid NAH7 are on a defective transposon. Mol Gen Genet 1990; 223: 33-9.
[PMID: 2175388]

[50]   Pieper DH, Martins dos Santos VAP, Golyshin PN. Genomic and mechanistic insights into the biodegradation of organic pollutants. Curr Opin Biotechnol 2004; 15: 215-24.
[PMID: 15193329]

[51]   Brazil GM, Kenefick L, Callanan M, *et al.* Construction of a rhizosphere pseudomonad with potential to degrade polychlorinated biphenyls and detection of bph gene expression in the rhizosphere. Appl Environ Microbiol 1995; 61: 1946-52.
[PMID: 7646029]

[52]   Kingsley MT, Fredrickson JK, Metting FB, *et al.* Environmental Restoration Using Plant-Microbe Bioaugmentation. Hinchee RE, Leeson A, Semprini L, *et al.* Eds Bio-remediation of Chlorinated and Polyaromatic Hydrocarbon Compounds. Boca Raton: Lewis Publishers 1994; pp. 287-92.

[53]   Sone Y, Mochizuki Y, Koizawa K, *et al.* Mercurial-resistance determinants in *Pseudomonas* strain K-62 plasmid pMR68. AMB Express 2013; 3: 41.
[http://dx.doi.org/10.1186/2191-0855-3-41] [PMID: 23890172]

[54]   Balamurugan D, Udayasooriyan C, Kamaladevi B. Chromium (VI) reduction by *Pseudomonas putida* and *Bacillus subtilis* isolated from contaminated soils. Int J Environ Sci 2014; 5: 522-9.

[55]   Jacobsen CS. Plant protection and rhizosphere colonization of barley by seed inoculating herbicide degrading bacteria *Burkholderia (Pseudomonas) cepacia* DBO1 (pRO101) in 2,4-D contaminated soil. Plant Soil 1997; 189: 139-44.

[56]   Schnoor JL, Aitchison EW, Kelley SL, *et al.* Phytoremediation and bioaugmentation of 1,4- dioxane. Abstract Adv Biol Rern Systematic Confer 1998; 87: 91-133.

[57]   Juhasz AL, Naidu R. Bioremediation of high molecular weight polycyclic aromatic hydrocarbons: a review of the microbial degradation of benzo [a] pyrene. Int Biodeterior Biodegradation 1996; 45: 57-88.

[58]   Mena-Benitez GL, Gandia-Herrero F, Graham S, *et al.* Engineering a catabolic pathway in plants for the degradation of 1,2-dichloroethane. Plant Physiol 2008; 147: 1192-8.
[PMID: 18467461]

[59]   Novotný C, Vyas BRM, Erbanová P, *et al.* Removal of PCBs by various white rot fungi in liquid cultures. Folia Microbiol (Praha) 1997; 42: 136-40.
[PMID: 9340310]

[60]   Liu L, Jiang CY, Liu XY, *et al.* Plant-microbe association for rhizoremediation of chloronitroaromatic pollutants with Comamonas sp. strain CNB-1. Environ Microbiol 2007; 9: 465-73.
[PMID: 17222144]

[61]   Bisht S, Pandey P, Sood A, *et al.* Biodegradation of naphthalene and anthracene by chemo-tactically active rhizobacteria of populus deltoides. Braz J Microbiol 2010; 41: 922-30.
[PMID: 24031572]

[62]   Bisht S, Pandey P, Aggarwal H, *et al.* Utilization of endophytic strain *Bacillus* sp. SBER3 for biodegradation of polyaromatic hydrocarbons (PAH) in soil model system. Eur J Soil Biol 2014; 60: 67-76.

[63]   Achal V, Pan X, Fu Q, *et al.* Biomineralization based remediation of As(III) contaminated soil by *Sporosarcina ginsengisoli.* J Hazard Mater 2012; 201-202: 178-84.
[PMID: 22154871]

[64]   Dong G, Wang Y, Gong L, *et al.* Formation of soluble Cr (III) end-products and nanoparticles during Cr (VI) reduction by *Bacillus cereus* strain XMCr-6. Biochem Eng J 2013; 70: 166-72.

[65]   Rahman A, Nahar N, Nawani NN, *et al.* Bioremediation of hexavalent chromium (VI) by a soil-borne bacterium, *Enterobacter cloacae* B2-DHA. J Environ Sci Health Part A Tox Hazard Subst Environ Eng 2015; 50: 1136-47.
[PMID: 26191988]

[66]   Kang JW. Removing environmental organic pollutants with bioremediation and phytoremediation. Biotechnol Lett 2014; 36: 1129-39.
[PMID: 24563299]

[67]   Dixit R, Malaviya D, Pandiyan K, *et al.* Bioremediation of heavy metals from soil and aquatic environment: an overview of principles and criteria of fundamental processes. Sustainability 2015; 7: 2189-212.

[68] Rojas LA, Yáñez C, González M, *et al.* Characterization of the metabolically modified heavy metal-resistant *Cupriavidus metallidurans* strain MSR33 generated for mercury bioremediation. PLoS One 2011; 6: e17555.
[http://dx.doi.org/10.1371/journal.pone.0017555] [PMID: 21423734]

[69] McLean KJ, Sabri M, Marshall KR, *et al.* Biodiversity of cytochrome P450 redox systems. Biochem Soc Trans 2005; 33: 796-801.
[PMID: 16042601]

[70] Van Aken B, Stahl JD, Naveau H, *et al.* Transformation of 2, 4, 6-trinitrotoluene (TNT) reduction products by lignin peroxidase (H8) from the white-rot basidiomycete *Phanerochaete chrysosporium.* Bioremediat J 2000; 4: 135-45.

[71] Kaplan O, Vejvoda V, Plíhal O, *et al.* Purification and characterization of a nitrilase from *Aspergillus niger* K10. Appl Microbiol Biotechnol 2006; 73: 567-75.
[PMID: 17061133]

[72] Kumar Ramasamy R, Congeevaram S, Thamaraiselvi K. Evaluation of isolated fungal strain from e-waste recycling facility for effective sorption of toxic heavy metal Pb (II) ions and fungal protein molecular characterization-a mycoremediation approach. Asian J Exp Biol Sci 2011; 2: 342-7.

[73] Achal V, Kumari D, Pan X. Bioremediation of chromium contaminated soil by a brown-rot fungus, *Gloeophyllum sepiarium.* Res J Microbiol 2011; 6: 166-71.

[74] Sukumar M. Reduction of hexavalent chromium by *Rhizopus oryzae.* Afr J Environ Sci Technol 2010; 4: 412-8.

[75] Menn FM, Easter JP, Sayler GS. Genetically Engineered Microorganisms and Bioremediation. In: Rehm HJ, Reed G, Eds. Biotechnology: Environmental Processes II. Germany, Weinheim: Wiley-VCH Verlag GmbH 2000; pp. 425-39.

[76] Gu JD, Mitchell R. Biodeterioration. In: Rosenberg E, DeLong EF, Lory S, Stackebrandt E, Thompson F, Eds. The Prokaryotes. Berlin, Heidelberg: Springer 2013; pp. 309-41.

[77] Enning D, Garrelfs J. Corrosion of iron by sulfate-reducing bacteria: new views of an old problem. Appl Environ Microbiol 2014; 80: 1226-36.
[PMID: 24317078]

[78] Sand W. Microbial mechanisms of deterioration of inorganic substrates-a general mechanistic overview. Int Biodeterior Biodegradation 1997; 40: 183-90.

[79] Goodell B, Qian Y, Jellison J. Fungal Decay of Wood: Soft Rot-Brown Rot-White Rot. In: Schultz TP, Militz H, Freeman MH, Eds. *et al.* , Development of Commercial Wood Preservatives – Efficacy, Environmental, and Health Issues. Wasington, DC, USA: American Chemical Society 2008; 982: pp. 9-31.

[80] Bhatnagar P, Khan AA, Jain SK, Rai MK. Biodeterioration of Archaeological Monuments and Approach for Restoration. In: Jain SK, Khan AA, Rai MK, Eds. Geomicrobiology. Boca Raton: CRC Press 2010; pp. 255-302.

[81] Srivastava S, Kumar R, Singh VP, Eds. Wood Decaying Fungi. Saarbrücken, Germany: LAP Lambert Academic Publishing 2013; p. 76.

[82] Pinheiro AC, Sequeira SO, Macedo MF. Fungi in archives, libraries, and museums: a review on paper conservation and human health. Crit Rev Microbiol 2019; 45: 686-700.
[PMID: 31815562]

[83] Orlita A. Microbial biodeterioration of leather and its control: a review. Int Biodeterior Biodegradation 2004; 53: 157-63.

[84] Szostak-Kotowa J. Biodeterioration of textiles. Int Biodeterior Biodegradation 2004; 53: 165-70.

[85] Ford T, Mitchell R. The Ecology of Microbial Corrosion. In: Marshall KC, Ed. Advances in Microbial

Ecology. New York: Plenum Press 1990; Vol. 11: pp. 231-62.

[86]   Jiang Z, Shi M, Shi L. Degradation of organic contaminants and steel corrosion by the dissimilatory metal-reducing microorganisms *Shewanella* and *Geobacter* spp. Int Biodeterior Biodegradation 2020; 147: 104842.
[http://dx.doi.org/10.1016/j.ibiod.2019.104842]

[87]   Kushkevych I. Isolation and Purification of Sulfate-Reducing Bacteria. Blumenberg M, Shaaban M, Elgaml A, Eds. Microorganisms. IntechOpen 2019; pp. 1-19.

[88]   Zhang G, Gong C, Gu J, *et al.* Biochemical reactions and mechanisms involved in the biodeterioration of stone world cultural heritage under the tropical climate conditions. Int Biodeterior Biodegradation 2019; 143: 104723.
[http://dx.doi.org/10.1016/j.ibiod.2019.104723]

[89]   Biswas J, Sharma K, Harris KK, *et al.* Biodeterioration agents: Bacterial and fungal diversity dwelling in or on the pre-historic rock-paints of Kabra-pahad, India. Iran J Microbiol 2013; 5: 309-14.
[PMID: 24475341]

[90]   Jurado V, Sanchez-Moral S, Saiz-Jimenez C. Entomogenous fungi and the conservation of the cultural heritage: A review. Int Biodeterior Biodegradation 2008; 62: 325-30.

[91]   Restrepo-Flórez J, Bassi A, Thompson MR. Microbial degradation and deterioration of polyethylene – a review. Int Biodeterior Biodegradation 2014; 88: 83-90.

[92]   Webb JS, Nixon M, Eastwood IM, *et al.* Fungal colonization and biodeterioration of plasticized polyvinyl chloride. Appl Environ Microbiol 2000; 66: 3194-200.
[PMID: 10919769]

[93]   Sequeira S, Cabrita EJ, Macedo MF. Antifungals on paper conservation: An overview. Int Biodeterior Biodegradation 2012; 74: 67-86.

[94]   Johnsrud SC. Biotechnology for solving slime problems in the pulp and paper industry. In: Eriksson K-EL, Babel W, Blanch HW, Eds. *et al.* , Biotechnology in the Pulp and Paper Industry Advances in Biochemical Engineering/Biotechnology. Berlin, Heidelberg: Springer 1997; 57: pp. 311-28.

[95]   Iverson WP. Microbial Corrosion of Metals. Laskin AI, Ed. Advances in Applied Microbiology. Academic Press 1987; Vol. 32: pp. 1-36.

CHAPTER 11

Azolla: A Better Prospective for Biological Nitrogen Fixation and Sustainable Agriculture in Era of Climate Change

Priyanka Chandra[1,*], **Parul Sundha**[1], **Rinki** [2], **Pooja Verma**[1], **Savitha Santosh**[3] and **Vanita Pandey**[2]

[1] *ICAR-Central Soil Salinity Research Institute, Karnal, 132001, Haryana, India*

[2] *ICAR-Indian Institute of Wheat and Barley Research, Karnal, 132001, Haryana, India*

[3] *ICAR- Central Institute for Cotton Research, Nagpur, Maharashtra - 440010, India*

Abstract: The use of nitrogen in an efficient way in agriculture has economic as well as environmental challenges. Bio-fertilizers and green manures are eco-friendly and economical sources for enhancing nitrogen use efficiency (NUE) for sustainable agriculture. In the era of climate change, conjunctive application of both bio-fertilizers and chemical fertilizers is required for soil health and sustainable yield as well. *Azolla* is one of the Biofertilizer that has the potential to fix nitrogen biologically, increase nitrogen recovery and enhance the rice yield. The regular application of *Azolla* significantly increases soil organic nitrogen content, which is much more beneficial than inorganic nitrogen. *Azolla* possesses the potential to mitigate major problems that are of global concern and can be used as a multi-faceted biofertilizer. Usage of *Azolla* in agriculture has various advantages as it has a positive impact on enhanced productivity and reduces input costs. They are also involved in the bioremediation of heavy metals and several toxic pollutants. Hence, it possesses great potential for its usage as a biofertilizer in the era of climate change.

Keywords: *Azolla*, Bio-fertilizers, Biological nitrogen fixation, Green manures, Nitrogen use efficiency.

INTRODUCTION

The efficient use of nitrogen (N) is the foremost requirement for agricultural sustainability as well as environmental security. Due to a lack of frugality in nitrogen use efficiency (NUE), more than $60–100$ Kg N yr^{-1} of reactive nitrogen gets dissipated into the environment every year, which leads to an adverse effect on the environment and human health [1]. There is an urgent need to monitor the

* **Corresponding author Priyanka Chandra:** ICAR-Central Soil Salinity Research Institute, Karnal, 132001, Haryana, India; E-mail: priyanka.chandra921@gmail.com

Ashutosh Gupta, Shampi Jain & Neeraj Verma (Eds.)

NUE in agriculture [2]. With an exponential increase in the world population, crop production needs to be doubled to accomplish food security.

Accordingly, the application of nitrogen fertilizers is expected to increase, which may cause environmental damage if the crop NUE is not substantially improved [1, 3, 4]. Therefore, the use of biofertilizers and green manures could be eco-friendly economically feasible, and scientific options for maintaining nitrogen requirements are some steps to sustain crop yields and maintain soil health in the era of climate change. *Azolla* can fix nitrogen up to 30–100 kg N ha^{-1} crop^{-1} and is very efficient in rice crops [5]. Several studies have demonstrated that even at a low concentration of nitrogen, *Azolla* has the potential to increase nitrogen recovery and enhance the rice yield [6] and regular application of *Azolla* significantly increases soil organic nitrogen content. Many research reports for the improvement of NUE by *Azolla* bio-fertilizer [7 - 9] are available.

AZOLLA-THE POTENTIAL BIO-FERTILIZER

Azolla is an aquatic fern that floats freely in fresh water and possesses the potential for biological nitrogen fixation through forming a symbiotic relationship with *Anabaena azollae*. *A. Azollae* is a cyanobacterium that resides in the dorsal lobes of the leaves of *Azolla*, potentially supplying a substantial amount of nitrogen to the rice crop [10, 11]. The name *Azolla* is derived from the Greek word *azo*, which means 'to dry' and *allyo* means 'to kill'. This means that a plant dies when it dries. J.B. Lamark, in 1973, established the genus *Azolla* and placed it in the order Salviniales and family Salvinaceae [12, 13]. *Azolla* is a small, fast-growing dichotomously branched aquatic fern, which is naturally available in places where water is available, such as moist soils, ditches, and marshy ponds, and hence distributed globally [14, 15]. The Indian species is generally triangular with a length of 1.5- 3.0 cm and a breadth of about 1- 2 cm. Rhizomes are the short, branched stem that are covered with leaves that overlap alternately and have two separated lobes, dorsal and ventral. The sporophyte of *Azolla* is organized in a dorsiventral manner [16, 17]. In the dorsal leaf lobe, there is an ellipsoidal cavity, and inside this cavity, *Anabaena Azollae* and *Arthrobacter,* Gram-positive bacteria reside [18].

Azolla has the potential to be used as green manure, as it provides a substantial amount of nitrogen to crops, especially rice crops. In rice crops, *Azolla* fixes nitrogen and provides nutrients to the plant that enhances crop growth. *Azolla* can be applied before the transplantation of rice or by intercropping with rice [19, 20]. *Azolla* has a very high multiplication rate as it can increase its biomass up to two-fold within 2 to 5 days under an optimum environment. *Azolla* can provide an almost partial amount of the requisite nitrogen in rice crops through nitrogen fixation. Along with nitrogen-fixing ability, *Azolla* has several other benefits too, which include weed suppression, reducing water temperature and pH, and lowering the volatilization of ammonia. *Azolla* also reduces the proliferation of mosquitoes.

Advantages of *Azolla* as a Biofertilizer over Chemical Fertilizers

1. *Azolla* is an environmentally friendly, natural Biofertilizer that converts atmospheric nitrogen into a utilizable form of nitrogen that can be absorbed by plants.

2. It also provides micronutrients, vitamins, and other growth substances that can be useful for plants and animals.

3. *Azolla* improves soil health, the structure of the soil, and its fertility by providing organic matter.

4. Several essential amino acids (nitrogen 4-5%, phosphorous 0.5-0.9%), vitamins (vitamin A, vitamin B_{12}, and beta-carotene), growth regulators, and minerals like calcium, phosphorous, potassium, ferrous, copper, zinc, and magnesium are found in *Azolla* [21 - 23]. On a dry weight basis, it was found that *Azolla* is composed mainly of protein (25–35%), minerals (10-15%), amino acids (7–10%), bioactive substances, and biopolymers [24].

There are several other beneficial properties of *Azolla* and its usage, like how it may minimize greenhouse gas mitigation and also help with soil and water management [25 - 27].

Beneficial Effects of *Azolla*

1. There are several reports on the beneficial properties of *Azolla,* indicating its role in improving soil fertility by increasing total nitrogen, organic carbon, and available phosphorus [28]. The *Azolla* green manure has a C/N ratio of between 9 and 10, which felicitates the release of NH_4^+ −N when applied to rice fields [29]. By acting as a physical barrier, *Azolla* can effectively reduce the NH_3 loss in the environment and improve the uptake of NH_4^+ by plants [30].
2. Comparative efficiency in the aspect of N and P availability made by *Azolla* species was studied by Singh and coworkers [31], and the results demonstrated that *A. Pinnata* was more efficient than *A. filiculoides* and *A. mexicana*. However, an Indian isolate of *A. Pinnata* was recorded with a maximum release of NH_4^+−N (88%), followed by a Vietnam isolate (77%) at 40 days of flooding. An increase in available P in soil from 18 ppm to 26 ppm was also reported because of the *Azolla* application. According to Xie and coworkers [32], the

most effective method to increase soil fertility is to culture *Azolla* as a monocrop separately and then incorporate it into rice growing fields. Kondo and coworkers [33] reported a 27% increase in the grain yield of rice as compared to urea when *A. Pinnata* was applied as an intercrop.

3. *Azolla* also provides organic matter to the soil, which improves the physical structure of the soil. It is also reported that *Azolla* decreases the specific gravity of soil and hence boosts soil porosity by up to 5% [3, 34]. In a study carried out by Awodun [35], *Azolla* was applied to the soil at a rate of 80 g kg^{-1}. The results demonstrated that soil bulk density was reduced to 1.34 g cm^{-3} from 1.44 g cm^{-3} in the control. Similarly, porosity increased 6.52% over control. *Azolla* also makes P and K available and exchangeable, respectively in the soil [36]. Several researchers also demonstrated that *Azolla* has the potential to better absorb potassium, under low-potassium conditions. In comparison to rice, *Azolla* has a greater absorption capacity for potassium as they absorb potassium in traces ranging from 0–5 ppm, while rice requires at least 8 ppm. Hence, *Azolla* can also be used as a good source of potassium, as it can supply potassium to crops after decomposition [37, 38]. *Azolla* can also make nutrients available to plants and enhance nutrient acquisition resulting in better growth and yield [39] (Fig. **1**).

Azolla also provides several advantages to the environment, along with plants, which are described below:

- Enhances crop growth and yield.
- Quality of the grains is improved.
- Reduction in the occurrence of diseases.
- Felicitate seedling establishment, hence improving their survival.
- Protect crops from biotic and abiotic stresses.
- Improves retention of water and enhances drought tolerance in crops.
- Provides balanced application of fertilizers use.
- Improves nitrogen use efficiency.
- Ameliorate salt stress by the accumulation of excess salt.
- Improves quality of the soil and its health.
- Enhance nutrient acquisition in crops.

Azolla can fix nitrogen up to 40-60 kg N ha^{-1} per rice crop and can enhance the yield of rice to 1.5–2.4 t ha^{-1} through fixation of 22–40 kg N ha^{-1} in a month and around 64–75 kg N ha^{-1} in two months [40, 41]. One of the advantages of *Azolla* is that it can be grown simultaneously with rice and does not require any additional costs for land or water [42].

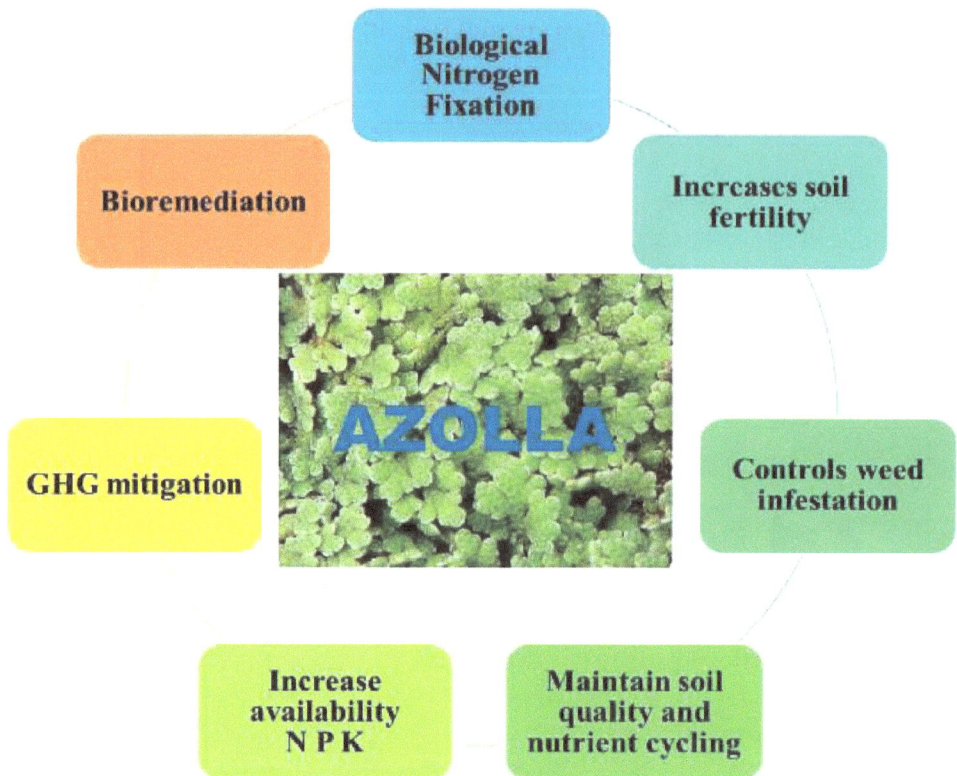

Fig. (1). Benefits of *Azolla*.

In addition, *Azolla* also reduces evaporation and transpiration in salt-affected areas, which further can be useful for reclaiming such soils and also controls weed infestation in rice crops [43]. *Azolla* can be useful in purifying wastewater as it can absorb phosphorus and heavy metals as well as water [44, 45]. The study carried out by Choudhury and coworkers [46] showed that *Azolla* bio-fertilizer in combination with urea could increase rice yield as well as improve nitrogen use efficiency and reduce nitrogen loss. *Azolla* bio-fertilizer improves nitrogen uptake in crops, reduces nitrogen loss in fields, and substantially increases nitrogen use efficiency.

However, there are many constraints on *Azolla* cultivation, which are described below:

- Managing water availability.
- Lack of phosphorus in soils.
- *Azolla* is attacked by several predators.
- Optimum temperatures are required for *Azolla* production.

- Improper management practices.
- Sufficient labour is required.
- Input costs for its production.

AZOLLA IN RICE CULTIVATION

Rice (*Oryza sativa*) is the major food crop that grows around the globe and is the staple crop for more than half of the world's population. Rice production is also increasing over the years along with its demands, which are specifically due to adaptations of scientific interventions and practices. Global paddy production in 2016 reached up to 751.9 million tonnes (499.2 million tonnes, milled basis) [47] and, China was the world's leading paddy producer, followed by India.

A high concentration of nutrients especially nitrogen is necessary for life cycle of rice [48, 49]. For the production of around one tonne of grain and straw, rice removes almost 16 kg of nitrogen from the soil [50], which results in a deficiency of nitrogen in the rice fields in almost all parts of the world. To overcome the deficiency of nitrogen, an increased amount of the nitrogen fertilizer is being applied to meet the crop demand. With increasing yield, the use of chemicals has increased. However, these chemicals are not being absorbed by plants. The nitrogen use efficiency is quite low and leads to environmental problems such as denitrification, ammonia volatilization, and leaching losses. All these nitrogen losses lead to the emission of greenhouse gases such as N_2O and NH_3 [51, 52]. Nitrogen left in residues from nitrates causes leaching in fields and toxicity in groundwater. The excessive use of urea also leads to the depletion of soil organic matter, which leads to a loss in fertility [53, 54]. Due to such problems, it has become necessary to increase the nitrogen use efficiency through changing the strategies for efficient use of nutrients and by integrating organic and eco-friendly fertilizers into rice production.

The use of organic products is one of the most important strategies and also seems particularly important [55, 56]. Agriculture also plays a significant role in the emission of greenhouse gases, particularly CO_2, especially from rice fields, which can be reduced by the use of organic products. Also, if soil stable organic matter can be increased, then agricultural soils can become powerful carbon sinks. In such conditions, *Azolla* as a bio-fertilizer can be a better option [57 - 59]. Due to its quick decomposition in soil and efficient nitrogen availability to rice, *Azolla* is becoming an alternative to commercial nitrogen fertilizers and is gaining popularity as a biofertilizer in several countries like the Philippines, Thailand, Srilanka, India and many other tropical areas [60 - 62]. The common species of *Azolla* that are being used as biofertilizers are *A. microphylla, A. filiculoides, A. pinnata, A. caroliniana, A. nilotica, A. rubra*, and *A. Mexicana* [63, 64]. Due to high multiplication rate, *Azolla* can suppresses diseases and weeds, and additionally, it decomposes very fast providing rich nutrients to the crops. Azolla's properties lend themselves to its use as green manure and biofertilizer for rice crops [65, 66]. It has been reported that *Azolla* produces around 300 tonnes of green bio-hectare per year under a normal subtropical climate, which is comparable to 800 Kg of nitrogen, *i.e.*, around 1800 Kg of urea [67, 68]. In the rice fields, basal application of green manure at the rate of 10-12 tonnes/hectare saves around 30-35 kg of nitrogen fertilizer required and increases nitrogen up to 50-60 kg/ha in the soil [69]. Usage of *Azolla* as an intercrop in rice at the rate of 500 kg/ha provides soil nitrogen up to 50 kg/ha along with saving around 20-30 kg/ha of nitrogen, which results in the enhancement of rice yield by up to 20 to 30%.

The properties of *Azolla*, which make it suitable for bio-fertilizer in rice, are described as follows:

1. *It* has the potential to fix atmospheric nitrogen.

2. It has a high multiplication rate.

3. *Azolla* does not compete with rice for light as it floats on the water surface and thus does not obstruct photosynthesis.

4. *Azolla* requires partial shade for its growth, and rice can provide a canopy in the required phonological growth stages. At later stages, low light and scarcity of nutrients lead to the death phase, which starts decomposition and releases nutrients into the soil.

5. Fast decomposition of *Azolla* makes nitrogen, phosphorus, and other nutrients available in the soil during grain development.

6. *Azolla* is a better potassium accumulator in a low potassium environment, and after its decomposition, the nutrients are available for rice.

7. *Azolla* rapidly increases its biomass and forms a thick mat, which suppresses weeds and diseases in rice fields.

METHODOLOGY FOR THE MASS MULTIPLICATION OF *AZOLLA*

The nursery method is the simplest method for *Azolla* multiplication on a large scale. This method can be easily adopted by farmers and can be carried out in fields. A 40-square-meter plot, farmyard manure (FYM), phosphorus, insecticides, and *Azolla* inoculum are all required materials. The procedure is as follows:

- Prepare the plot of 20 x 2m and then irrigate it and maintain the water level at a height of 10 cm.
- Then make the slurry using 10 kg of FYM with 20 ls of water and spray it on the plots.
- Thereafter, add phosphorus to it.
- Inoculate each plot with 8 kg of fresh *Azolla* culture.
- After *Azolla* inoculation, add phosphorus to it again on the 4th and 8th days at 100g.
- After the 7th day of *Azolla* inoculation, spray with insecticide at the rate of 100 g/plot.
- Throughout the incubation period of two to three weeks, maintain the water level at the initial (10 cm) height.
- After incubation, *Azolla* can be harvested by collecting and draining extra water.

PROCEDURE OF *AZOLLA* APPLICATION IN RICE CROPS

There are two methods of application of *Azolla* as a bio-fertilizer:

1. Green manure: *Azolla* is applied before transplanting rice, and fresh *Azolla* biomass is directly applied to fields. However, this methodology requires lots of fresh *Azolla*.

2. Duel cropping: In this method, *Azolla* is applied after transplanting rice directly to the field. *Azolla* can be grown as a dual culture with rice and incorporated subsequently.

CONCLUDING REMARKS

Azolla possesses the potential to mitigate major problems that are of global concern and can be used as a multi-faceted biofertilizer. Adaptation of *Azolla* in agriculture not only increases productivity but also improves sustainability in the era of climate change. It also reduces agricultural input costs, which has a positive impact on socio-economic factors. *Azolla* has a positive impact on environmental issues as it reduces the usage of chemical nitrogen and protects the environment from the side effects of residual nitrogen. The nitrogen fixation ability of *Azolla* fixes atmospheric nitrogen and also sequesters atmospheric CO_2, so emissions of greenhouse gases such as CH_4 and N_2O can be reduced. Hence, it can be concluded that *Azolla* technology provides an option for economic as well as sustainable agriculture.

CONSENT FOR PUBLICATION

Not applicable.

CONFLICT OF INTEREST

The author declares no conflict of interest, financial or otherwise.

ACKNOWLEDGEMENTS

Declared none.

REFERENCES

[1]   Lassaletta L, Billen G, Grizzetti B, *et al.* 50 year trends in nitrogen use efficiency of world cropping systems: the relationship between yield and nitrogen input to cropland. Environ Res Lett 2014; 9: 105011.
        [http://dx.doi.org/10.1088/1748-9326/9/10/105011]

[2]   Steffen W, Richardson K, Rockström J, *et al.* Sustainability. Planetary boundaries: guiding human development on a changing planet. Science 2015; 347: 1259855.
        [http://dx.doi.org/10.1126/science.1259855] [PMID: 25592418]

[3]   Yao Y, Zhang M, Tian Y, *et al. Azolla* biofertilizer for improving low nitrogen use efficiency in an intensive rice cropping system. Field Crops Res 2018; 216: 158-64.

[4]   Hirel B, Tétu T, Lea PJ, *et al.* Improving nitrogen use efficiency in crops for sustainable agriculture. Sustainability 2011; 3: 1452-85.

[5]   Roy D, Pakhira M, Bera S. A review on biology, cultivation and utilization of *Azolla.* Adv Life Sci 2016; 5: 11-5.

[6]   Singh AL, Singh PK, Lata P. Effects of different levels of chemical nitrogen (urea) on *Azolla* and blue-green algae intercropping with rice. Fert Res 1988; 17: 47-59.

[7]   Jampeetong A, Sripakdee T, Khamphaya T, *et al.* The effects of nitrogen as NO_3^- and NH_4^+ on the growth and symbiont (*Anabaena azollae*) of *Azolla pinnata* R. Brown. Chiang Mai Univ J Nat Sci 2016; 15: 11-20.

[8]   Watanabe I, Padre BJr, Ramirez C. Mineralization of *Azolla* N and its availability to wetland rice. Soil Sci Plant Nutr 1991; 37: 679-88.

[9]   Mian MH, Kashem MA. Comparative efficiency of some selected methods of applying *Azolla* for cultivation of irrigated rice. Bangladesh J Crop Sci 1995; 6: 29-36.

[10]   Bhuvaneshwari K, Singh PK. Response of nitrogen-fixing water fern Azolla biofertilization to rice crop. 3Biotech 2015; 5: 523-9.

[11]   Parashuramulu S, Swain PS, Nagalakshmi D. Protein fractionation and *in vitro* digestibility of *Azolla* in ruminants. J Appl Anim Res 2013; 3: 129-32.

[12]   Svenson HK. The new world species of *Azolla.* Am Fern J 1944; 34: 69-84.

[13]   Konar RN, Kapoor RK. Anatomical studies on *Azolla pinnata.* Phytomorphol 1972; 22: 211-23.

[14]   Peters GA, Mayne BC. The *Azolla, Anabaena azollae* relationship: I. Initial characterization of the association. Plant Physiol 1974; 53: 813-9.
        [PMID: 16658796]

[15]   Peters GA, Meeks JC. The *Azolla-Anabaena* symbiosis: basic biology. Annu Rev Plant Physiol Plant Mol Biol 1989; 40: 193-210.

[16]   Plazinski J, Zheng Q, Taylor R, *et al.* DNA probes show genetic variation in cyanobacterial symbionts of the *azolla* fern and a closer relationship to free-living nostoc strains than to free-living anabaena strains. Appl Environ Microbiol 1990; 56: 1263-70.
        [PMID: 16348182]

[17]   Neilan BA. Identification and phylogenetic analysis of toxigenic cyanobacteria by multiplex randomly amplified polymorphic DNA PCR. Appl Environ Microbiol 1995; 61: 2286-91.
[PMID: 16535049]

[18]   Brouwer P, Bräutigam A, Buijs VA, *et al.* Metabolic adaptation, a specialized leaf organ structure and vascular responses to diurnal N_2 fixation by *Nostoc azollae* sustain the astonishing productivity of *Azolla* ferns without nitrogen fertilizer. Front Plant Sci 2017; 8: 442.
[http://dx.doi.org/10.3389/fpls.2017.00442] [PMID: 28408911]

[19]   Kollah B, Patra AK, Mohanty SR. Aquatic microphylla *Azolla*: a perspective paradigm for sustainable agriculture, environment and global climate change. Environ Sci Pollut Res Int 2016; 23: 4358-69.
[PMID: 26697861]

[20]   Yadav RK, Abraham G, Singh YV, *et al.* Advancement in the utilization of *Azolla-Anabaena* system in relation to sustainable agricultural practices. Proc Indian Natn Sci Acad 2014; 80: 301-16.

[21]   Pillai PK, Premalatha S, Rajamony S. *Azolla*: a sustainable feed for livestock. LEISA India 2002; 4: 15-7.

[22]   Chichilichi B, Mohanty GP, Mishra SK, *et al.* Effect of partial supplementation of sun-dried *Azolla* as a protein source on the immunity and antioxidant status of commercial broilers. Vet World 2015; 8: 1126-30.
[PMID: 27047208]

[23]   Islam MA, Nishibori M. Use of multivitamin, acidifier and *Azolla* in the diet of broiler chickens. Asian-Australas J Anim Sci 2017; 30: 683-9.
[PMID: 27507178]

[24]   Fasakin EA. Nutrient quality of leaf protein concentrates produced from water fern (*Azolla africana* Desv) and duckweed (*Spirodela polyrrhiza* L. Schleiden). Bioresour Technol 1999; 69: 185-7.

[25]   Rai LC, Kumar HD, Mohn FH, *et al.* Services of algae to the environment. J Microbiol Biotechnol 2000; 10: 119-36.

[26]   Prasanna R, Kumar V, Kumar S, *et al.* Methane production in rice soil is inhibited by cyanobacteria. Microbiol Res 2002; 157: 1-6.
[PMID: 11911608]

[27]   Rai PK. Wastewater management through biomass of *Azolla pinnata*: an eco-sustainable approach. Ambio 2007; 36: 426-8.
[PMID: 17847810]

[28]   Syamsiyah J, Sunarminto BH. Mujiyo. Changes in soil chemical properties of organic paddy field with *Azolla* application. J Soil Sci Agroclimat 2016; 13: 68-73.

[29]   Watanabe I, Lapis MT, Oliveros R, Ventura W. Improvement of phosphate fertilizer application to *Azolla*. Soil Sci Plant Nutr 1988; 34: 557-69.

[30]   Ram H, Krishna RP, Naidu MVS. Effect of *Azolla* on soil properties and yield of mungbean (*Vigna radiate* L.). J Indian Soc Soil Sci 1994; 42: 385-7.

[31]   Singh PK, Panigrahi BC, Satapathy KB. Comparative efficiency of *Azolla*, blue-green algae and other organic manures in relation to N and P availability in a flooded rice soil. Plant Soil 1981; 62: 35-44.

[32]   Xie J, Wu X, Tang J, *et al.* Chemical fertilizer reduction and soil fertility maintenance in rice-fish coculture system. Front Agric China 2010; 4: 422-9.

[33]   Kondo M, Kobayashi M, Takahashi E. Effect of phosphorus on *Azolla* and its utilization in rice culture in Niger. Plant Soil 1989; 120: 165-70.

[34]   Subedi P, Shrestha J. Improving soil fertility through *Azolla* application in low land rice: a review. Azarian J Agric 2015; 2: 35-9.

[35]   Awodun MA. Effect of *Azolla* (*Azolla* species) on physicochemical properties of the soil. World J

Agric Sci 2008; 4: 157-60.

[36]   Kumar B, Shahi DK. Effect of *Azolla* as green manure on soil properties and grain yield of rice in acid soil of Jharkhand. Ann Plant Soil Res 2016; 18: 214-8.

[37]   Dawar S, Singh PK. Growth, nitrogen fixation and occurrence of epiphytic algae at different pH in the cultures of two species of *Azolla.* Biol Fertil Soils 2001; 34: 210-4.

[38]   Sudadi S. *Azolla* based organic farming: low biotechnology for high rice productivity. IJPAES 2014; 4: 425-9.

[39]   Ali S, Watanabe I. Response of *Azolla* to P, K, and Zn in different wetland rice soils in relation to chemistry of flood water. Soil Sci Plant Nutr 1986; 32: 239-53.

[40]   Peoples MB, Herridge DF, Ladha JK. Biological nitrogen fixation: an efficient source of nitrogen for sustainable agricultural production. Plant Soil 1995; 174: 3-28.

[41]   Cisse M, Vlek PLG. Influence of urea on biological N_2 fixation and N transfer from *Azolla* intercropped with rice. Plant Soil 2003; 250: 105-12.

[42]   Singh A, Singh PK. Intercropping of *Azolla* biofertilizer with rice at different crop geometry. Trop Agric 1990; 67: 350-4.

[43]   Krock T, Alkamper J, Watanabe I. *Azolla* contribution to weed control in rice cultivation. Plant Syst Evol 1991; 34: 117-29.

[44]   Shiomi N, Kitoh S. Nutrient absorption capacity of *Azolla* from wastewater and use of *Azolla* plant as biomass. J Plant Nutr 1987; 10: 1663-70.

[45]   Miao XL, Wu QY, Yang CY. Fast pyrolysis of microalgae to produce renewable fuels. J Anal Appl Pyrolysis 2004; 71: 855-63.

[46]   Choudhury ATMA, Kecskes ML, Kennedy IR. Utilization of BNF technology supplementing urea N for sustainable rice production. J Plant Nutr 2014; 37: 1627-47.

[47]   FAO. Rice Market Monitor 2017; 20: 41. [https://www.fao.org/3/cb6989en/cb6989en.pdf]

[48]   Roger PA, Ladha JK. Biological N_2 fixation in wetland rice fields: estimation and contribution to nitrogen balance. Plant Soil 1992; 141: 41-5.

[49]   Roger PA, Watanabe I. Technologies for utilizing biological nitrogen fixation in wetland rice: potentials, current usage and limiting factors. Fert Res 1986; 9: 39-77.

[50]   Akhter S, Mian MH, Kaderm MA, *et al.* Combination of *Azolla* and urea nitrogen for satisfactory production of irrigated boro rice (BRRI Dhan 29). J Agron 2002; 1: 127-30.

[51]   Mahapatra BS, Sharma GL. Integrated management of Sesbania, Azolla and urea nitrogen in low land rice under a rice-wheat cropping system. J Agric Sci 1989; 113: 203-26.

[52]   Perreira AL, Martins M, Oliviera MM, *et al.* Morphological and genetic diversity of the family Azollaceae inferred from vegetative characters and RAPD markers. Plant Syst Evol 2011; 297: 213-26.

[53]   De Datta SK, Ed. Principle and Practises Rice Production. NY, USA: John Wiley & Sons 1981; p. 642.

[54]   Choudhury ATMA, Kennedy IR. Prospects and potential for systems of biological nitrogen fixation in sustainable rice production. Biol Fertil Soils 2004; 39: 219-27.

[55]   Van Coppenolle B, Watanabe I, Van Hove C, *et al.* Genetic diversity and phylogeny analysis of Azolla based on DNA amplification by arbitrary primers. Genome 1993; 36: 686-93.
       [PMID: 8405985]

[56]   Watanabe I, Berja NS, Del Rosario DC. Growth of *Azolla* in paddy field soils as effected by phosphorus fertilizer. Soil Sci Plant Nutr 1980; 26: 301-7.

[57]   Jain SK, Vasudevan P, Jha NK. Removal of some heavy metals from polluted water by aquatic plants: studies on duckweed and water valvet. Biol Wastes 1989; 28: 115-26.

[58]   Cohen-Shoel N, Barkay Z, Ilzycer D, *et al.* Bio-filtration of toxic elements by *Azolla* biomass. Water Air Soil Pollut 2002; 135: 93-104.

[59]   Arora A, Sexana S, Sharma DK. Tolerance of phytoaccumulation of chromium by three *Azolla* species. World J Microbiol Biochem 2006; 22: 97-100.

[60]   Jeyapandiyan N, Lakshmanan A. Influence of *Azolla* and blue green algae in residual nitrogen and uptake in different rice cultivation systems. Trends Biosci 2014; 7: 1504-6.

[61]   Joshi LN, Rana A, Shivay YS. Evaluating the potential of rhizocyanobacteria as inoculants for rice and wheat. Agric Technol Thail 2012; 8: 157-71.

[62]   Raja W, Rathaur P, John SA, *et al. Azolla*: an aquatic pteridophyte with great potential. Inter J Res Biol Sci 2012; 2: 68-72.

[63]   Habte M. *Azolla filiculoides* nitrogenase activity decrease induced by inoculation with *Chlamydomonas* sp. Appl Environ Microbiol 1986; 52: 1123-7. [PMID: 16347211]

[64]   Kannaiyan S, Kumar K, Eds. Biodiversity of Azolla and its algal symbiont Anabaena azollae. Chennai, TN, India: National Biodiversity Authority 2006; p. 17.

[65]   Bocchi S, Malgioglio A. *Azolla-Anabaena* as a biofertilizer for rice paddy fields in the Po Valley, a temperate rice area in northern Italy. Int J Agron 2010; 2010: 152158. [http://dx.doi.org/10.1155/2010/152158]

[66]   Mazid M, Khan TA. Future of bio-fertilizers in Indian agriculture: an overview. Intern J Agric Food Res 2014; 3: 10-23.

[67]   Roger PA, Kulasooriya SA, Eds. Blue-Green Algae and Rice. Los Banos, Philippines: The International Rice Research Institute 1980; p. 112.

[68]   Arora A, Singh PK. Comparison of biomass productivity and nitrogen fixing potential of *Azolla* spp. Biomass Bioenergy 2003; 24: 175-8.

[69]   Rivaie AA, Isnaini S. Maryati. Changes in soil N, P, K, rice growth and yield following the application of *Azolla pinnata.* J Biol Agric Healthc 2013; 3: 112-8.

CHAPTER 12

Management of Plant-parasitic Nematodes Infesting Pomegranate and Guava

B.S. Chandrawat[1], Harshraj Kanwar[2], Rashid Pervez[3,*] and A.U. Siddiqui[4]

[1] *SKN College of Agriculture, SKNAU, Jobner (Rajasthan), India*

[2] *Rajasthan Agricultural Research Institute, Durgapura, SKNAU, Jobner (Rajasthan), India*

[3] *Division of Nematology, ICAR-Indian Agricultural Research Institute, New Delhi, India*

[4] *Pacific College of Agriculture, Pacific University, Udaipur (Rajasthan), India.*

Abstract: Pomegranate and guavas are two important commercial fruit crops grown in India. These fruit crop plants can be attacked by insect pests, plant pathogens, and plant-parasitic nematodes, which can reduce the quality and quantity of fruit production. Diseases caused by plant-parasitic nematodes are of great economic importance. Many species of plant-parasitic nematodes are found associated with pomegranates and guavas. Root-knot nematodes, *Meloidogyne incognita* and *M. javanica* are among the economically important nematode pests of pomegranates in the world as well as India. In India, these nematodes are a serious problem in Maharashtra state. Guava orchards are facing symptoms of sudden decline and loss in productivity due to heavy infestation of a highly pathogenic species of root-knot nematode, *M. enterolobii*. The root-knot nematode is spreading to new areas through infected pomegranate and guava saplings. Management of this nematode is a challenge because of its polyphagous nature and ability to survive on weed hosts. Different scientific management strategies are discussed in this chapter.

Keywords: Disease, Guava, Management, Nematodes, Pomegranate.

INTRODUCTION

India is the second-largest producer of fruit in the world. Different varieties of fruits are grown in different agro-climatic zones of India. Pomegranates and guavas are important commercial fruit crops grown in India. Such an important fruit crop plant can be attacked by insect pests, plant pathogens, and plant-parasitic nematodes, which can reduce the quality and quantity of fruit production [1, 2]. Diseases caused by plant-parasitic nematodes are of great economic importance. Many species of plant-parasitic nematodes are found associated with

* **Corresponding author Rashid Pervez:** Division of Nematology, ICAR-Indian Agricultural Research Institute, New Delhi, India; E-mail: rashidpervez2003@gmail.com

Ashutosh Gupta, Shampi Jain & Neeraj Verma (Eds.)

major pomegranate and guava growing areas of India [3, 4] and some of these species are highly damaging to pomegranate and guava production [3, 5]. Hashim [6] has studied parasitic nematodes associated with pomegranates in Jordan.

Darekar and coworkers [3] reported ten plant-parasitic nematodes associated with pomegranates in Maharashtra, India, whereas Khan and coworkers [7] reported thirteen species of eight plant-parasitic nematode genera on guava. *Pratylenchus* sp., *Meloidogyne* sp., and *Rotylenchulus reniformis* are widely known as important plant parasites, which can be of concern for the decline in yield over the years. In addition, they indirectly aggravate soil-borne pathogens, particularly fungi and bacteria, leading to wilt problems. Fanelli and coworkers [8] recovered two different nematode species from decaying pomegranate fruit in Southern Italy. El-Qurashi and coworkers [9] reported a 98% infestation of the root-knot nematode on pomegranates in the Assiut governorate. Twelve nematode genera were recorded from the rhizosphere of pomegranates in Balochistan province [10] (Table **1**).

Table 1. Diversity of plant-parasitic nematodes associated with pomegranate and guava.

Crop	Nematodes Specie	Infested Locality	Reference(s)
Pomegranate	*Aphelenchus avenae*	Balochistan	[10]
	Criconemella antipolitana	Jordan	[6]
	Criconemoides sp.	Hebron, Palestine	[17]
	Criconemoid xenoplax	Jordan	[6]
	Ditylenchus sp.	Balochistan	[10]
	Helicotylenchus sp.	Hebron, Palestine	[17]
	H. digonicus	Hebron, Palestine	[17]
		Balochistan	[10]
	H. dihystera	Himachal Pradesh	[18]
	H. minzi	Jordan	[6]
	H. pseudorobustus	Jordan	[6]
	Hoplolaimus indicus	Balochistan	[10]
	Longidorus sp.	Jordan	[6]
		Balochistan	[10]
	Meloidogyne sp.	Jordan	[6]
	Meloidogyne javanica	Egypt	[13]
		Balochistan	[10]

(Table 1) cont.....

Crop	Nematodes Specie	Infested Locality	Reference(s)
	Meloidogyne incognita	Egypt	[13]
		Balochistan	[10]
		Himachal Pradesh	[18]
		Peru	[14]
	Merlinius brevidens	Balochistan	[10]
	Paratrichodorus tunisiensis	Jordan	[6]
	Pratylenchus sp.	Hebron, Palestine	[17]
	P. penetrans	Jordan	[6]
	Psilenchus hilarulus	Balochistan	[10]
	Panagrellus sp.	Italy	[8]
	Scutylenchus rugosus	Balochistan	[10]
	Sheraphelenchus sucus	Italy	[8]
	Tylenchorhynchus sp.	Hebron, Palestine	[17]
	T. brassicae	Balochistan	[10]
	T. clarus	Jordan	[6]
	Tylenchus sp.	Balochistan	[10]
	Xiphinema sp.	Hebron, Palestine	[17]
	X. basiri	Balochistan	[10]
	X. index	Jordan	[6]
	X. pachtaicum	Jordan	[6]
		Hebron, Palestine	[17]

(Table 1) cont.....

Crop	Nematodes Specie	Infested Locality	Reference(s)
Guava	*Aphelenchus avenae*	West Bengal	[7]
	Helicotylenchus sp.	Colombia	[19]
	H. abunamai	West Bengal	[7]
	H. goodi	West Bengal	[7]
	H. indicus	West Bengal	[7]
		New Delhi	[20]
	Hemicriconemoides strictathecatus	New Delhi	[20]
	Hoplolaimus indicus	Mediterranean Region	[21]
		India	[22]
		West Bengal	[7]
		New Delhi	[20]
	Meloidogyne spp.	Malaysia	[23]
		Colombia	[19]
	M. enterolobii	Puerto	[24]
		Africa	[25]
		Europe	[26]
		United States	[27]
		Vietnam	[28]
		China	[29]
		Thailand	[30]
		Uttarakhand	[31]
		Karnataka	[32]
	M. incognita	Uttar Pradesh	[33]
		West Bengal	[7]
	M. javanica	West Bengal	[7]
	M. graminicola	West Bengal	[7]
	Mesocriconema sphaerocephala	New Delhi	[20]
	Pratylenchus sp.	Malaysia	[23]
		Colombia	[19]
	P. brachyurus	West Bengal	[7]
	P. coffeae	West Bengal	[7]
	Radopholus similis	India	[22]
	Rotylenchulus reniformis	India	[22]
		West Bengal	[7]
		New Delhi	[20]
	Tylenchorhynchus sp.	United Arab Republic	[34]

(Table 1) cont.....

Crop	Nematodes Specie	Infested Locality	Reference(s)
	T. mashhoodi	West Bengal	[7]
		New Delhi	[20]
	T. nudus	West Bengal	[7]
	Tylenchulus semipenetrans	New Delhi	[20]

Root-knot nematodes, *Meloidogyne incognita,* and *M. javanica* are the world's most economically important parasites of pomegranate cultivars [11, 12]. They have also been reported from northern Egypt and Peru [13, 14].

M. enterolobii has been reported in several countries growing guava. It was first reported from Aubergine in Puerto Rico and later from Africa, Europe, the United States, Vietnam, China, and Thailand (Table **1**). Poornima and coworkers [15] reported *M. enterolobii* for the first time from India. The nematode was detected in guava orchards showing a sudden decline in Tamil Nadu. Singh [16] reported that guava orchards are experiencing symptoms of sudden decline and loss in productivity due to heavy infestation of highly pathogenic species of root-knot nematode, *M. enterolobii,* which not only causes a heavy loss by predisposing the host to secondary attack by the wilt pathogenic fungus *Fusarium oxysporum* f. sp. *psidii,* causing a disease complex with synergistic effects on guava. Most of the fruit crops are grown as monocrops that provide a congenial environment to the nematodes. If the nematode population is left unchecked, it will increase to an alarming level and may eventually wipe out the whole crop. Nematode management is therefore important for high yields, and quality that is required by the high cost of modern crop production. The major fruit-growing states are facing a remarkable loss due to nematode infestation (Table **2**).

Table 2. Yield loss by plant-parasitic nematodes in pomegranate and guava.

Crop	Nematodes	Yield Losses (%)	Reference
Pomegranate	*Meloidogyne* spp.	24.6-27.4	[35]
	M. incognita	31.17	[36]
	M. incognita	32	[37]
	Meloidogyne spp.	17.3	[12]
Guava	*M. enterolobii*	60-80	National Institute of Plant Health Management (NIPHM) data base
	M. incognita	23.77	[38]

The maximum number of nematodes is present at a distance of 25 to 50 cm from the base of the plant and at a depth of 20 to 40 cm in the soil. However, nematodes also live when crop roots get deep into the soil [12]. According to a study by Ganesh and Sharma [18], a 30 x 15 cm distance depth is most appropriate for plant-parasitic nematode in the rhizosphere of up to 10 years old pomegranate plants. Total nematode populations were found inversely proportional to the sampling distance and depth in the rhizosphere, *i.e.*, with the increase in the distance and depth, nematode populations were found to be reduced. In pomegranate orchards, a depth of 15 cm revealed higher populations of plant-parasitic nematodes.

The nematode itself cannot move very far in the soil. It can be easily spread by any outside force that moves soil, such as farm equipment, irrigation, or heavy rainfall. The nematode can easily spread on plant materials such as sweet potatoes, storage roots, or ornamental plants. Root-knot nematode is spreading to new areas through infected pomegranate and guava saplings. Management of this nematode is a challenge because of its polyphagous nature and ability to survive on weed hosts. Even the *Mi* gene, which confers natural resistance to root-knot nematodes, proved ineffective against *M. enterolobii*.

SYMPTOMS

Nematodes are minute microorganisms that live in the soil and mostly feed on plant roots. Low nematode populations in soil cause damage to plant roots, but disease symptoms do not appear on the upper part. When a high population occurs in the soil, the infected plants show disease symptoms. The disease symptoms that appear on above-ground plant parts are similar to nutrient deficiency symptoms and other common diseases. Identification of nematode disease symptoms is very difficult.

The most commonly associated symptoms observed were stunting of plants, the small size of leaves and fruits, and yellowing of leaves. The intensity of root-knot nematode damage increased with the increase in the age of the plant. The soil type has a profound effect on nematode reproduction. It is highest in coarsely textured sandy soils and least in more finely textured soils. Soil type, moisture, and temperature enhance the degree of root-knot nematode infestation and population size.

Severely infected trees decline rapidly, the leaves mature early, and they defoliate during dry periods and when stressed by water, leaving a few clusters of pale yellow leaves at the apical end of the branches. Severely stunted $2^{1/2}$-year-old guava plants could be easily pulled out of the ground, revealing the swollen, ugly, and discoloured root system. Moderate infestations by nematodes are associated with general chlorosis, nutrient deficiency symptoms, and reduced flowering and fruiting. The effect of the nematodes on the roots results in stunted growth of the above-ground parts of the plants. The trees bore small, pale green to yellowish chlorotic foliage. Plants infected by nematodes exhibit poor growth. The symptoms start with the yellowing of plants followed by atrophy, giving the tree a parched look and revealing dirty roots with spangled knots. Plants become flaccid when broken, revealing a hollow branchlet. The quality and quantity of fruits produced by nematode-infested trees were also affected. The younger plants tended to produce a lot of young fruits, which aborted readily. The fruits lacked a smooth, glossy, green appearance. Fruits that developed to maturity matured early, were smaller in size, and became more susceptible to fungal infection. Similar quality fruits were also found on older bearing trees, which survived the heavy infestation of the nematodes. Fruits collected from the nematode-infected plants are low-weighed (Fig. **1**).

Fig. (1). (A) Nematode-infested guava plant in the field, **(B)** Root-knot disease in guava root, and **(C)** Nematode-infested guava plant in the nursery.

The nematode-infested plant manifested a drastic reduction in plant growth, small leaves, an absence of fine roots, a poorly developed root system, and a decline in yield quality and quantity. Severely infested roots are distorted by small and large multiple galls.

Banihashemi [39] reported a decline in pomegranate production mostly in newly established 3-7 year old orchards. The first disease symptoms usually appear in early summer as one side yellowing, wilting, and death of some branches or the entire tree. The collar region of the diseased plants contains loose bark with flaky tissues. The roots of such trees are healthy and capable of producing vigorous suckers. Decline symptoms were usually observed less frequently on older trees.

Nematodes associated with pomegranate trees cause severe damage to the thickness of the roots and stunting. Lesions were formed at the site of entry with necrosis, enlargement of cells into giant cells containing egg masses, and sections of larvae.

The incidence of nematode disease in the guava field was first noted on 6-month old plants. On 2 to 4 year old plants the manifestation of symptoms was distinct. Roots of root-knot nematode infected trees show multiple galls and root necrosis, causing a drastic decrease in fine roots. The primary and secondary lateral roots, distorted by the large multiple galls, were dark brown or black, with rough corky surfaces and cracks along the long axes of the roots. There were distinct reductions in the root systems, which were almost devoid of fine feeder roots. Young, white roots emerging from the swollen infected tissues were also infested by the nematodes.

The genus *Meloidogyne* [40] causes damage in the form of root galls and a reduction in the number of roots, and predisposition to fungal (*e.g.*, *Fusarium*, *Phytophthora*, *Rhizoctonia*, and *Verticillium*) and bacterial diseases further aggravates the damage and cause losses in crop yields [41, 2]. Young galls are white and change to brown and are hardy when they become old. This is due to hypertrophy and hyperplasia of root cortical cells and is partly due to the formation of giant cells on which the nematodes feed. The gall formation results in the disturbance of the normal transmission of nutrients from the roots to the shoot. The root-knot nematode infection decreases the efficiency of roots in the uptake of water and nutrients from the soil. Furthermore, root-knot nematodes often thrive and cause damage to perennial hosts for many years, preventing them from reaching their full yield potential. These are specialized endoparasitic and lead parasitic modes of life. Luxuriant green trees with no flowering for a long period of time with their large root-knot are also the symptoms of severe nematode infestation.

The decline and death of guava trees parasitized by *M. enterolobii* is a complex disease involving other microorganisms like *Fusarium solani* [42], *Verticillium dahliae*, and *Pythium aphanidermatum* [43]. This complex disease is causing serious economic problems for growers and the economy.

MANAGEMENT

Management in the Nursery

Use of Soilless Culture Medium

Coco pit, coir pith, or vermiculite can be used to prepare the planting material.

Use of Nematode-Free Soil

Preparation of nematode-free planting material in a nursery is a good option using sterilized soil. The soil can be sterilized by several methods.

Soil Solarization

The use of thin transparent polythene sheets (100 gauges) during peak summer (2-3 months) has been found to be very effective and may be used in the nursery area. Alternatively, use solarized soil for filling the polythene bags before establishing the grafts from mother plants into the polythene bags or nursery.

Summer Ploughing

Deep summer ploughing of the nursery/main field before establishing or transplanting of seedlings.

Air Layering

It may be a safe method of vegetative propagation. The rooting media used is soilless to ensure protection and then shifted to sterilized media for further growth of plants.

Treatment of Soil Mixture Used for Raising Rootstocks or Planting Materials

A tonne of soil mixture mixed with 50-100 kg of oil-cake enriched with bioagents such as *Paecilomyces lilacinus* (1 kg), *Pseudomonas fluorescens* (1 kg), *Trichoderma harzianum* or *T. viride*(1 kg) can also be used.

Management in Orchards

Site Preparation Before Planting for Nematode Management

The site selected for establishing a new orchard should not have a history of vegetables as a previous crop, as they may leave an inoculum of root-knot nematode, reniform nematode, or lesion nematode.

Collection of Soil Samples from the Proposed Site for Nematode Population Estimation

Based on the soil sample report for nematodes, further field preparation measures can be taken. If a nematode problem is seen in the sample, then before transplanting, use precautionary measures for nematode management, *i.e.*, use deep summer ploughing, keep the land free from any plants and weeds.

Repeat sampling after 2-3 months for nematode population estimation till the site is free from major plant-parasitic nematodes, and it should be repeated every year for better management of nematodes and to gain quality production.

Management in Field

- Crop rotation in the main field before transplanting the crop.
- Intercropping with sunhemp, marigold, mustard, fenugreek, onion, garlic, *etc.* during the fallow/rest period.
- Application of FYM/compost at 20-25 t/ha or non-edible oil cakes at 2 t/ha in the nursery as well as in the main field at transplanting and at 'Bahar' (blossom) every year.
- Application of bioagents (*Trichoderma viride, P. lilacinus, Pseudomonas fluorescens, etc.*) at 10 to 20 g/m^2 nursery or at 5 kg with 100 kg of farmyard manure (FYM)/compost/ha in the main field at transplanting as well as at 'Bahar' (blossom) every year.
- At 3–4 month intervals, apply 3–4 kg of bio-agent enriched vermicompost/FYM/compost per plant.
- Soil application with carbofuran 3G or phorate 10G at 2 kg a.i./ha in the nursery as well as at transplanting in the main field and at 4 kg a.i./ha at 'Bahar' (blossom) every year.
- Akhter and Khan [44] found nematicidal or nematostatic properties in aqueous leaf extracts of *Xanthium strumarium* L. (Cocklebur), *Acalypha indica* (Indian Mercury), *Colocasia gigantea* (Indian taro), and *Argemone mexicana* (Mexican poppy) against *M. incognita in vitro*. These can also be applied in the field.
- Application of *T. Harzianum* in the root zones of guava plants reduced the number of nematodes in both soil and roots as compared to the number of untreated plants. Moreover, inoculation of guava plants with *T. harzianum* arrested the development of the juvenile nematodes [30].
- Application of carbofuran 10G was the most effective treatment in suppressing the nematode densities on guava and fig trees. The most effective management combinations, next to carbofuran 10G, in suppressing the nematode densities in the rhizosphere of guava trees were *P. lilacinum* + *P. penetrans* + urea (46%), *P. lilacinum* + *P. penetrans* + chicken manure, and *T. harzianum* + *P. penetrans* + chicken manure (66.54–69.22% nematode reductions) [45].

Cultural Practices for the Management of Nematodes

- The movement of nematode-infected rootstocks should be strictly restricted.
- Use nematode-free planting material or avoid infested seedlings for planting.
- Bare-root seedlings can be immersed in 45°C hot water for 25 minutes.
- Removal and destruction of nematode-infested planting material or trees.

- Maintaining the orchard free from alternate hosts, *i.e.*, weeds and susceptible crops.

CONCLUDING REMARKS

The root-knot nematodes *M. incognita* and *M. javanica* are the most economically important nematode pests of pomegranates in the world. In India, these nematodes are a serious problem in Maharashtra, India. Guava orchards are facing symptoms of sudden decline and loss in productivity due to heavy infestation of a highly pathogenic species of root-knot nematode, *M. enterolobii*. Adoption of integrated pest management in the field, as well as nursery schedules, is important in these crops since excessive use of pesticides could lead to pesticide residues in the produce, affecting human health and causing other ecological hazards.

CONSENT FOR PUBLICATION

Not applicable.

CONFLICT OF INTEREST

The author declares no conflict of interest, financial or otherwise.

ACKNOWLEDGEMENTS

The authors expressed their gratitude to the Director and Joint Director (Research), ICAR-Indian Agricultural Research Institute, New Delhi, and Dean, SKN College of Agriculture, SKNAU, Jobner (Rajasthan), India.

REFERENCES

[1]   Dias-Arieira CR, Furlanetto C, Santana SM, *et al.* Fitonematoidesassociados a frutíferasnaregiãonoroeste do Paraná, Brasil. Rev Bras Frutic 2010; 32: 1064-71.

[2]   Sikora RA, Coyne D, Hallmann J, *et al.* 3rd Eds. Plant-parasitic Nematodes in Subtropical and Tropical Agriculture. Wallingford, Oxfordshire: CABI Publishing, 2018; p. 898.

[3]   Darekar KS, Shelke SS, Mhase NL. Nematode associated with fruit crops in Maharashtra. Int Nematol Network Newslett 1990; 7: 11-2.

[4]   Salalia R, Siddiqui AU, Parihar A. Two new species of *Xiphinema* Cobb, (Nematoda: Dorylamida) associated with perennials in Udaipur with notes on a known associated species. Indian J Nematol 2002; 32: 58-62.

[5]   Shelke SS, Darekar KS. Reaction of pomegranate germplasm to root-knot nematode. J Maharashtra Agric Univ 2001; 25: 308-10.

[6]   Hashim Z. Plant-parasitic nematodes associated with pomegranate *Punica granatum* in Jordan and an attempt to chemical control. J Nematologia Med 1983; 11: 199-200.

[7]   Khan MR, Hassan A, Ghosh B, *et al.* Diversity and community analyses of soil nematodes associated with guava from West Bengal, India. Acta Hortic 2007; 735: 483-8.

[8]   Fanelli E, Troccoli A, Vovlas N, *et al.* Occurrence of *Sheraphelenchus sucus* (Nematoda:

Aphelenchoidinae) and *Panagrellus* spp. (Rhabditida: Panagrolaimidae) associated with decaying pomegranate fruit in Italy. J Nematol 2017; 49: 418-26.
[PMID: 29353931]

[9]   El-Qurashi MA, El-Zawahry AMI, Abd-El-Moneem KMH, *et al.* Occurrence, population density and biological control of root-knot nematode, *Meloidogyne javanica* infecting pomegranate orchards in Assiut Governorate Egypt. Assiut J Agric Sci 2019; 50: 176-89.

[10]   Khan A, Shaukat SS, Siddiqui IA. A survey of nematodes of pomegranate orchards in Balochistan Province, Pakistan. Nematolmedit 2005; 33: 25-8.

[11]   Chitwood BG. Root-knot nematodes. Part-I A revision of the genus *Meloidogyne* Goeldi, 1887. Proceed Helminthological Soc 1949; 16: 90-104.

[12]   Singh T, Prajapati A, Maru AK, *et al.* Root-knot nematodes (*Meloidogyne* spp.) infecting pomegranate: a review. Agric Rev (Karnal) 2019; 40: 309-13.

[13]   Ibrahim IKA, Mokbel AA. Occurrence and distribution of the root-knot nematodes *Meloidogyne* spp. and their host plants in northern Egypt. Egypt J Exp Biol (Botany) 2009; 5: 125-9.

[14]   Vega-Callo RA, Mendoza-Lima MY, Mamani-Mendoza NR, *et al.* First report of southern root-knot nematode, *Meloidogyne incognita*, infecting pomegranate, *Punica granatum*, in Peru. J Nematol 2020; 52: 1-3.
[PMID: 32329969]

[15]   Poornima K, Suresh P, Kalaiarasan P, *et al.* Root-knot nematode, *Meloidogyne enterolobii* in guava (*Psidium guajava* L.) a new record from India. Madras Agric J 2016; 103: 359-65.

[16]   Singh N. Emerging problem of guava decline caused by *Meloidogyne enterolobii* and *Fusarium oxysporum* f. sp. *psidii.* Indian Phytopathol 2020; 73: 373-4.

[17]   Wahbeh AJA, Ed. Survey of plant-parasitic nematodes associated with the rhizosphere of plants in nurseries and intensive agriculture in the west bank. Hebron, Palestine: Hebron University 2007; p. 107.

[18]   Ganesh F, Sharma GC. Spatial distribution of nematodes in pomegranate orchard. Indian J Nematol 2017; 47: 179-92.

[19]   Piedrahita G, Adrian O, Jairo CZ. Identification of plant-parasitic nematodes in guava (*Psidium guajava* L.), at the municipality of Manizales (Caldas), Colombia. Rev Acad Colomb Cienc Exactas Fis Nat 2010; 34: 117-26.

[20]   Gasti V, Chawla G, Rao U, *et al.* Community analysis of nematodes infecting guava and mango at IARI and the first report of citrus nematode from guava orchard. Indian J Nematol 2015; 45: 195-201.

[21]   Lamberti F, Vovlas N. Plant-parasitic nematodes associated with olive. Bull OEPP 1993; 23: 481-8.

[22]   Siddiqui MA. Further studies for the integrated management of the stunt, lance and spiral nematodes with cropping sequences ploughing. Arch Phytopathol Plant Protec 2003; 36: 9-22.

[23]   Razak AR, Lim TK. Occurrence of the root-knot nematode *Meloidogyne incognita* on guava in Malaysia. Pertanika 1987; 10: 265-70.

[24]   Rammah A, Hirschmann H. *Meloidogyne mayaguensis* n. sp. (Meloidogynidae), a root-knot nematode from Puerto Rico. J Nematol 1988; 20(1): 58-69.
[PMID: 19290185]

[25]   Willers P. First record of *Meloidogyne mayaguensis* Rammah and Hirschmann, 1988: Heteroderidae on commercial crops in the Mpumalanga province, South Africa. Inligtings bulletin-Instituut vir Tropiese en Subtropiese Gewasse 1997; 294: 19-20.

[26]   Blok VC. Mitochondrial DNA differences distinguishing *Meloidogyne mayaguensis* from the major species of tropical root-knot nematodes. Nematology 2002; 4: 773-81.

[27]   Brito J, Powers TO, Mullin PG, *et al.* Morphological and molecular characterization of *Meloidogyne*

mayaguensis isolates from Florida. J Nematol 2004; 36: 232-40.
[PMID: 19262811]

[28]   Iwahori H, Truc NTN, Ban DV, *et al.* First report of root-knot nematode *Meloidogyne enterolobii* on guava in Vietnam. Plant Dis 2009; 93: 675-5.
[PMID: 30764405]

[29]   Zhuo K, Hu MX, Liao JL, *et al.* First report of *Meloidogyne enterolobii* on arrowroot in China. Plant Dis 2010; 94: 271.
[PMID: 30754277]

[30]   Jindapunnapat K, Chinnasri B, Kwankuae S. Biological control of root-knot nematodes *(Meloidogyne enterolobii)* in guava by the fungus *Trichoderma harzianum.* J Dev Sustain Agric 2013; 8: 110-8.

[31]   Kumar S, Rawat S. First report on the root-knot nematode *Meloidogyne enterolobii* (Yang and Eisenback, 1988) infecting guava *(Psidium guajava)* in Udham Singh Nagar of Uttarakhand, India. Int J Curr Microbiol Appl Sci 2018; 7: 1720-4.

[32]   Ravichandra NG. New Report of root-knot nematode *(Meloidogyne enterolobii)* on guava from Karnataka, India. EC Agric 2019; 5: 504-6.

[33]   Ansari RA, Khan TA. Parasitic association of root-knot nematode, *Meloidogyne incognita* on guava. e-J Sci Tech (e-JST) 2012; 5: 65-7.

[34]   Diab KA, El-Eraki S. Plant-parasitic nematodes associated with olive decline in the United Arab Republic. J Plant Dis 1968; 52: 150-4.

[35]   Singh RV, Midha SK, Kumar V. Major Nematode Pests of Economically Important Crops in India. In: Trivedi PC, Ed. Jodhpur, India: Scientific Publishers 2003; pp. 233-56.

[36]   Walunj AR, Ed. Seasonal incidence and biological management of root- knot nematode, *Meloidogyne incognita* (Kofoid and White, 1919) Chitwood, 1949 infesting pomegranate, *Punica granatum* L. Rahuri: Mahatma Phule Krishi Vidyapeeth 2013; p. 103.

[37]   Khan MR, Jain RK, Ghule TM, *et al.* Root-knot nematodes in India-a comprehensive monograph All India Coordinated Research Project on Plant-parasitic nematodes with Integrated approach for their Control. New Delhi: Indian Agricultural Research Institute 2014; p. 78.

[38]   Patil SC, Mhase NL, Kulkarni SR. Assessment of avoidable yield losses due to *M. incognita* infesting guava under field conditions. Adv Life Sci 2016; 5: 6782-5.

[39]   Banihashemi Z. Etiology of pomegranate decline in Fars Province of Iran. Phytopath Med 1998; 37: 127-32.

[40]   Göldi EA. Relatóriosôbre a molestia do cafeirona provincial da Rio de Janeiro. Archuivos do Museu Nacional do Rio de Janeiro 1887; 8: 7-123.

[41]   Karssen G, Ed. The Plant-Parasitic Nematode Genus *Meloidogyne* Göldi, 1892 (Tylenchida) in Europe. Leiden. The Netherlands: Koninklijke Brill NV 2002; p. 160.

[42]   Gomes VM, Souza RM, Corrêa FM, *et al.* Management of *Meloidogyne mayaguensis* in commercial guava orchards with chemical fertilization and organic amendments. Nematol Bras 2010; 34: 23-30.

[43]   Avelar-Mejía JJ, Téliz-Ortiz D, Zavaleta-Mejía E. Patógenosasociados con el "declinamiento del guayabo". Rev Mex Fitopatol 2001; 19: 223-9.

[44]   Akhter G, Khan TA. Evaluation of some plant extracts for nemato-toxic potential against juveniles of *Meloidogyne incognita in vitro.* J Phytopharmacol 2018; 7: 141-5.

[45]   Dawabah AAM, Al-Yahya FA, Lafi HA. Integrated management of plant-parasitic nematodes on guava and fig trees under tropical field conditions. Egypt J Biol Pest Control 2019; 29: 29.
[http://dx.doi.org/10.1186/s41938-019-0133-9]

CHAPTER 13

Nematode Pests of Plantation Crops

Rashid Pervez[1,*] and **Raj Kumar**[2]

[1] *Division of Nematology, ICAR-Indian Agricultural Research Institute, New Delhi, India*

[2] *Division of Crop Protection, ICAR-CPCRI, Kasaragod, Kerala, India*

Abstract: The coconut palms grown in homestead and plantation situations suffer considerable damage due to infestation by plant-parasitic nematodes right from the seedling stage. Many plant feeder nematodes have been reported from coconut. Economically, the most important key nematode pest in India is the burrowing nematode, *Radopholus similis*, a migratory endoparasite causing typical root lesions and rotting symptoms on tender roots. An extensive, widespread occurrence was recorded in Kerala, Tamil Nadu, and Karnataka, causing a 30% yield loss in the coconut cropping system. Root-knot nematode, *Meloidogyne* spp., the devastating nematode pest of vegetables, spices, medicinal and fruit intercrop/mixed crops under coconut cropping system, also caused significant yield loss. This chapter discussed symptoms caused by nematode pests and different management practices.

Keywords: *M. incognita*, Management, Nematodes, Plantation crops, *Radopholus similis*.

INTRODUCTION

For several reasons, agriculture production in India is at a loss. Among them, one of the major constraints is diseases, which are the limiting factor in the cultivation of crops. Among them, diseases caused by plant-parasitic nematodes (PPN) are one of the constraints on reducing both the quality and yield of the crops. They cause 21.3% of crop losses, amounting to 102039.79 million ($1577 million) annually; the losses in nineteen horticultural crops were assessed at 50224.98 million, while for eleven field crops, it was estimated at 51814.81 million. The rice-rot nematode *Meloidogyne graminicola* is the most economically important, causing a yield loss of $2,372.32 million in rice. Citrus (9828.22 million) and banana (9710.46 million) were among the fruit crops that suffered the most casualties, while tomato (6035.2 million), brinjal (3499.12 million), and okra (2480.86 million) were among the vegetable crops that suffered the most losses

[*] **Corresponding author Rashid Pervez:** Division of Nematology, ICAR-Indian Agricultural Research Institute, New Delhi, India; E-mail: rashidpervez2003@gmail.com

Ashutosh Gupta, Shampi Jain & Neeraj Verma (Eds.)

[1, 2]. It has been estimated that global losses amount to $78 billion due to root-knot nematode (RKN) [3]. The coconut palms grown under homestead and plantation situations suffer considerable damage due to infestation by plant-parasitic nematodes right from the seedling stage.

More than 75 nematode species have been reported infecting coconut. Economically, the most important key nematode pest is in India, the burrowing nematode, *Radopholus similis*, is a migratory endoparasite causing typical root lesions and rotting symptoms on tender roots. The root-knot nematode, *Meloidogyne* spp. devastating nematode pest on vegetables, spices, medicinal and fruit intercrop/mixed crops under the coconut cropping system also cause significant yield loss. Red ring disease of coconut, caused by *Rhadinaphelenchus cocophilus* is a major problem faced by the coconut industry in Caribbean, Central, and South America. However, presently the disease does not have any reports on its occurrence in India.

THE BURROWING NEMATODE (*RADOPHOLUS SIMILIS*)

In plantation crops, *R. similis* is one of the most important root pathogens attacking coconut seedlings, with higher damage compared to adult palms. An extensive, widespread occurrence was recorded in Kerala, Tamil Nadu, and Karnataka and caused a 30% yield loss in the coconut cropping system. Other preferred hosts of this nematode are bananas, coconut, citrus, areca nuts, ginger, tea, and black pepper.

Damage Symptoms

- Nematodes burrow root tips, and their continuous feeding causes small, elongated orange-colored lesions on tender (white, creamy) roots, as well as lesions along with the crack formation on main roots. The lesions coalesce and lead to extensive rotting and cavities on the roots.
- Lesions are not conspicuous on secondary and tertiary roots as they are narrow and rot quickly on the infestation.
- Tender roots of coconut seedlings under heavy infestation become spongy in texture.
- Parasitized plants exhibited severe yellowing, stunting, and a reduction in the length and width of leaflets.

Management

- Use nematode-free healthy seedlings in the main field, which could perform better in the field and reduce the spread of the nematode through planting material.

- Maintaining field sanitation in coconut gardens by removing alternate hosts such as weeds and applying *Hydnocarpus* oil cake @ 4 kg/palm in June-July and October-November or neem cake @ 1 kg/pit is effective in suppressing nematode multiplication.
- Application of biological agents such as *Pasteuria penetrans*, *Paecilomyces lilacinus*, *Trichoderma*, and VAM (*Glomus fasciculatum*) has shown suppression of nematode multiplication.

ROOT-KNOT NEMATODE (*MELOIDOGYNE* SPP.)

It is the most common nematode pest in okra, tomato, cucurbits, amaranth, turmeric, black pepper, papaya, and non-cultivating intercrops or mixed crops in coconut gardens. Root galls are noticed if plants are infested by root-knot nematode (RKN). Such damage to root tissues leads to secondary infection by fungi and bacteria, either independently or together, causing complex diseases.

Damage Symptoms

Above Ground Symptoms

- First detected in localized areas within a field.
- Infested plants show stunting chlorosis (yellowing) of lower leaves, and plants may wilt during the heat of the day, especially under dry conditions or in sandier soils (Fig. **1A**).

Below-Ground Symptoms on Roots

- Infested roots look like swollen and distorted roots. Root-knot galls may vary in size and shape. On heavily infested plants, galls tend to fuse so that large areas, or the entire root, maybe swollen (Fig. **1B**).

Fig. (1). (A) A nematode infested coconut seedling and **(B)** Nematode infested coconut roots.

Management

Sanitation

The risks of introducing new nematodes can be reduced by purchasing or growing nematode-free transplants. If a crop is no longer valued in a garden, remove as much of the root system as possible to prevent a further build-up of nematode populations. Since nematodes may reproduce on certain weeds, weed control may also help to prevent an unnecessary increase in the nematode population. Seedlings or plants infested with nematodes noticed during field operations should be destroyed.

Crop Rotation

Rotating with a non-host crop will help reduce the RKN population to below damaging levels. Sorghum and maize planted for 2 to 3 consecutive years can provide excellent control of root-knot nematode. These crops are coming up well under coconut shade in the coastal sandy soils of Kasaragod, India, and will provide additional income to the farmers as well as green fodder for dairy animals.

Resistant Varieties

The use of resistant varieties is the most effective way to suppress nematode multiplication.

Nematode-Suppressive Crops

Nematode suppressive crops such as marigold (*Tagetes erecta*) have been found suitable for cultivation in the coconut cropping system. Such crops reduce nematode populations in the soil by depriving them of food or by releasing chemicals into the soil that inhibit their reproduction and development.

Organic Amendments

Application of neem cake at 150 g/sq. metre in the turmeric-coconut cropping system during August-September significantly suppresses the root-knot nematode population.

Biological Control

Application of *Trichoderma harzianum* (CP-28) enriched with neem cake at 1.0 kg/vine during the pre-monsoon and post-monsoon seasons significantly suppresses the root-knot nematode (*Meloidogyne incognita*) population and incidence of *Phytophthora* wilt in the black pepper-coconut cropping system.

Chemical Control

Many farmers may not wish to use this method of control because of economic or environmental concerns. However, in a very severe nematode problem, the application of carbosulfan at 1ml/liter (water drenching) to infested soils will bring down the population within a short period of treatment. After nematicide application, a light irrigation may also be given to ensure adequate soil moisture.

CONCLUDING REMARKS

In plantation crops, *R. similis* is one of the most important root pathogens, attacking coconut seedlings with higher damage compared to adult palms. An extensive, widespread occurrence was recorded in Kerala, Tamil Nadu, and Karnataka and caused a 30% yield loss in the coconut cropping system. Adoption of an integrated pest management schedule is important for these crops. Excessive pesticide use may result in pesticide residues in produce, endangering human health and posing other environmental risks.

CONSENT FOR PUBLICATION

Not applicable.

CONFLICT OF INTEREST

The author declares no conflict of interest, financial or otherwise.

ACKNOWLEDGEMENT

The authors expressed their gratitude to the Director and Joint Director (Research), ICAR-Indian Agricultural Research Institute, New Delhi, and thanks to the Director, ICAR-Central Plantation Crops Research Institute, Kasaragod (Kerala), India.

REFERENCES

[1]   Walia RK, Chakrabarty PK, Eds. Nematode Problems of Crops in India; A Comparative Volume on Four Decade of AICRP (Nematode) ICAR-All India Coordinated Research Project on Nematodes in agriculture. New Delhi: MS Printers 2018; p. 400.

[2]   Khan MR, Jain RK, Singh RV, *et al*. Eds. Economically Important Plant-parasitic Nematodes Distribution ATLAS Directorate of Information and Publications of Agriculture, Krishi Anusandhan Bhavan 1. New Delhi: Pusa 2010; p. 137.

[3]   Chen ZX, Chen SY, Dickson DW, Eds. Nematology Advance and Perspectives-Nematode Management and Utilization. UK: CAB International 2004; Vol. 2: p. 597.

CHAPTER 14

Nanotechnology Mediated Soil Microorganisms

F. Ahmad[1] and **S. Ahmad**[1,*]

JC Bose University of Science and Technology, YMCA, Faridabad, Haryana, India

Abstract: Biochemical interactions of nanoparticulate materials in the environment present a fairly complex situation due to a large number of available biochemical pathways. Insufficient knowledge about the interaction mechanisms involved means most of the experimental observations gathered are mixed up with ambiguous results. Taking the example of nanotechnology-enabled agriculture in the future, several beneficial impacts of green chemistry-based nanoparticulates (NPs) are expected to improve disease-tolerant crops with better yields. The critical issues involved in designing a plan of action in this context are briefly introduced in the present chapter after describing the agricultural bioorganisms and nanoparticulate species entering industrial plants on a large scale. This chapter aims to excite the imaginations of the readers by contributing to the future development of nanoagriculture.

Keywords: Importance, Nanoparticulate material, NP-soil microorganism interaction, Properties, Soil microorganisms.

INTRODUCTION

The increasing world population and the shortage of food are inevitable phenomena, and if adequate efforts are not made to make agriculture sustainable, the gap between food availability and food shortage will keep growing. This means more malnutrition, hunger, and poverty will spread fast. With agriculture being a major component of supporting the food cycle, it becomes pertinent to examine how to exploit the full potential of the soil nutrients available for improving agricultural production. However, the extraction of the soil nutrients is not easy as diverse location conditions prevail throughout. This perennial problem of an increasing population needing more food demands continuous efforts to exploit the use of soil nutrients to enhance food availability. Because of a complex interplay between plants taking the soil nutrients *via* soil microbes, better exploitation of these processes needs a further in-depth understanding of the participating mechanisms causing the molecular-level interactions.

* **Corresponding author S. Ahmad:** JC Bose University of Science and Technology, YMCA Faridabad, Haryana, India; E-mail: drsahmad@email.com

Ashutosh Gupta, Shampi Jain & Neeraj Verma (Eds.)

With the advent of nanotechnology in the recent past, the improved physicochemical properties of nanoparticles (NPs) of inorganic, organic, polymeric, and biomolecular origins have been witnessed in recent studies to identify the clear interaction pathways after exposing these nanoparticulate species to the crops, where the transport of lower dimension material species is becoming accessible for maximizing the intake of soil nutrients from the soil *via* soil microbes.

Agricultural farming, according to NIFA, US, is aimed at protecting the environment along with extending the utility of natural resources by maintaining and improving the soil fertility for different crops in different regions. This kind of sustainable agriculture would ultimately enhance the quality of life as a whole and make the farm profitable by enhancing agricultural production. Sustainable agriculture would, thus, involve an integrated system of plant and animal production practices using conventional and organic farming. Long-term goals of having sufficient human food, feed, fiber, and fuel production, would be necessary to meet the requirements of the rising global population combined with environmental protection and extended uses of natural resources [1].

In this context of exploring vast complexities inherited from the entire ecosystem and integrating them for enhanced and sustainable plant and animal production, it has caught the attention of researchers to examine the complex interaction of the crop plants with available soil nutrients by taking into account the route *via* numerous pathways involving a large variety of soil microbes.

The importance of having access to soil-borne nutrients to improve crop health is, in principle, amenable to control after knowing the metabolic activities of the soil microbes. Once these metabolic pathways of plant-soil microbe nutritional interactions are identified, it would be possible to implement programs to have better sustainable agricultural systems. However, the complex interplay between the plant microbiome and its modulation according to the plant's nutritional requirements has been at the root of meaningfully predicting the correct protocols, but, to date, most of the information is rather scattered. Despite the significant challenges in this context, some researchers have expressed optimism about incorporating the associated fragments of knowledge gathered to understand the mechanisms involved in governing the dynamic plant-microbe interactions [2]. The current attempts made to understand the significant influence of crop plants on taxonomy have already highlighted the need to identify the plant genes and the processes that control the way plants decide the rhizospheric microbiome. In particular, the deciphering of the plant genes that form the composition of the root exudates and participate in the metabolic pathways is influenced by the associated signals in the rhizosphere. To accomplish this objective, it is, therefore, necessary to collect and analyze the root exudates to identify the molecules that are used by the plants for communicating with soil microbes, along with the signal decryption of the pathways involved therein [2].

Nanoparticulate material species, which have been explored exhaustively in the recent past in drug and gene deliveries along with their theranostic applications, can also be used in the targeted delivery of fertilizers, pesticides, herbicides, plant growth regulators, and other formulations. Various advantages derived from the nanocarrier formulations are expected to offer better storage and precisely controlled release capabilities *via* different routes of encapsulation and entrapment involving polymers and dendrimers, surface ionic and van der Waal's attachments. These strategies are expected to enable the delivery of cargo and the onslaught of environmental degradation, which will reduce the amount to be applied. These surface-functionalized nanocarriers can be specially engineered to anchor the plant roots, surrounding soil structure, and organic matter. The success of these strategies is dependent on a thorough understanding of the molecular and conformational mechanisms at work between the delivery nanocarriers and the targeted molecular structures in the soil [3 - 5].

Current studies are directed at understanding the basic interaction mechanisms of NPs with the environment, including air, water, and soil, in which these NPs are expected to undergo considerable physico-chemico-biological modifications. This chapter is an example of using the novel physico-chemico-biological properties of NPs to control their interaction with a large variety of soil microorganisms to achieve improved and sustainable crop production, resulting in enhanced food production to combat the hunger problem globally.

The whole discussion in this chapter is divided into three parts, mainly: the introduction of the soil microorganisms, nanoparticulate materials and their physicochemical properties, and the interactions of NPs with soil microorganisms, resulting in useful as well as harmful effects. Instead of going into the details of the topics involved, efforts have been made to introduce the subject along with the proof of concepts with the latest published experimental results.

SOIL MICROORGANISMS: BRIEF INTRODUCTION

The soil microorganisms contain a variety of species, *viz.* bacteria, actinomycetes, fungi, algae, protozoa, and viruses loaded with different functional features in the soil. The study of soil microbiology identifies these organisms in the soil and explores their contributions, affecting the overall soil properties. These organisms, interacting among themselves, are found to influence the overall soil fertility.

Bacteria

Unicellular decomposer organisms, eat dead plant materials and organic wastes and release the nutrients by converting them from inaccessible to usable forms through the nitrogen cycle.

Actinomycetes

The actinomycetes are involved in providing the characteristic link between the bacteria and fungi and give soil its characteristic smell, besides being the source of several significant therapeutic medicines.

Fungi

These are the fibrous strings called hyphae that are 1 mm wide but several metres in length. Fungi promote nutrient breakdown and release into the soil. Fungi promote plant growth by attaching to plant roots through mycorrhizal interactions. In this way, the fungi provide nutrients to the plants and, in return, get carbohydrates from them. On the other hand, fungi that attach themselves to plants or other organisms get food as parasites. The fungi in the soil act as decomposers (saprophytic fungi), converting dead organic material into fungal biomass, carbon dioxide (CO_2), and organic acids; mycorrhizal fungi colonize the plant roots to obtain carbon from the plant. Mycorrhizal fungi help in making the phosphorus soluble and bring soil nutrients (phosphorus, nitrogen, micronutrients, and water) to the plant. One major group of mycorrhizae, namely ectomycorrhiza, grows on the surface layers of the roots of trees. Endomycorrhiza, the second major group of mycorrhizae, grows within the root cells and is commonly associated with grasses, row crops, vegetables, and shrubs. Parasites are the third group of fungi, pathogens or parasites, that cause reduced production or death when they colonize roots and other organisms.

Algae

These species are present in most soils in the presence of moisture and sunlight, and one gram of soil may vary from 100 to 10,000. They participate in photosynthesis and get CO_2 from the atmosphere and energy from sunlight for synthesizing their food. The major roles and functions of algae in soil include soil fertility maintenance, adding organic carbon to the soil after they die, preventing soil erosion by binding soil particles together, promoting water retention in the soil for longer periods, causing submerged aeration of oxygen in the soil *via* photosynthesis, checking nitrate losses through leaching and drainage in uncropped soils, and promoting rock weathering and soil structure formation.

Protozoa

These are colourless, unicellular animal-like organisms larger than bacteria, measuring a few microns to a few millimeters. Their population in a gram of arable soil may vary from 10,000 to 100,000, and they are abundant in surface soil. They withstand adverse soil conditions as they are characterized by a protected, dormant stage in their life cycle. The major functions, roles, and features of protozoa include getting nutrition from ingesting soil bacteria, besides maintaining the microbial/bacterial equilibrium and acting as biological control agents against organisms that cause harmful diseases to plants as well as human beings that are carried through water and other vectors.

Viruses

Viruses are involved in influencing the ecology of biological communities through transferring genes from host to host and as a potential cause of microbial mortality. Consequently, they influence the turnover and concentration of nutrients and gases. Despite its significant importance, the subject of soil virology needs more detailed investigation. Studies are currently being conducted into the virus diversity and abundance in different ecosystems to explore the role of viruses in plant health and soil quality. Viruses are highly abundant in all the areas studied so far, even in circumstances where bacterial populations differ significantly in the same environments. Soils probably harbour many novel viral species that, together, may represent a large reservoir of genetic diversity. According to some researchers, investigating this largely unexplored diversity of soil viruses has the potential to transform our understanding of the role of viruses in global ecosystem processes and the evolution of microbial life itself.

Nematodes

Nematodes are worms that are typically 50μm in diameter and about 1 mm in length. Though a number of species causing plant diseases have received much attention, far less is known about much of the nematode community that plays beneficial roles in the soil. An incredible variety of nematodes have been discovered functioning at several levels of the soil-food web. Some of them feed on plants and algae (the first level), others are grazers feeding on bacteria and fungi (second level), and some feed on other nematodes (higher levels). Nematodes have been divided into four broad groups based on their diets. Bacterial feeders consume bacteria. Fungal feeders are fed by puncturing the cell walls of fungi and sucking out the internal contents. Predatory nematodes eat all types of nematodes and protozoa, including smaller organisms whole or attach themselves to the cuticle of larger nematodes, scraping away until the entire internal body parts are extracted. Nematodes are found to release nutrients from the plants. When nematodes eat bacteria or fungi, ammonium is released because bacteria and fungi contain much more nitrogen than the nematodes require. Nematodes may also be useful indicators of soil quality due to their tremendous diversity and their participation in many functions at different levels of the soil-food web.

SOIL MICROORGANISMS: FUNCTIONAL BEHAVIOUR

Collectively, soil microorganisms decompose the organic matter, cycle the nutrients, and fertilize the soil. Without the cycling of elements, the continuation of life on Earth would be impossible, since essential nutrients would be rapidly taken up by organisms and locked in a form that could not be used by others. The reactions causing elemental cycling are often of chemical nature, but the biochemical reactions facilitated by organisms are also important and are assisted by the soil microbes. Soil microbes are thus responsible for the development of healthy soil. Soil microbes produce polysaccharides and mucilage that aid in cementing soil aggregates and make the aggregates less likely to crumble when exposed to water. Fungal filaments produce thread like structural branching in the soil surrounding particles and aggregates like a hairnet, promoting the stability of the soil structure. Thus, fungi act as the "threads" of the soil fabric. Microbes, in general, have little influence on the physical structure of soil, which is specifically done by larger organisms. Soil microorganisms are both components and producers of soil organic carbon locked into the soil for long periods of time, resulting in improved soil fertility and water-retaining capacity. Soil microorganisms and fungi may be able to draw carbon from the atmosphere and sequester it in the soil, reducing atmospheric greenhouse gases and limiting the impact of greenhouse gas-induced climate change. Healthy soils containing microbes and microbial biomass have enormous potential for improving the soil conditions from available carbon, moisture, aeration, temperature, acidity/alkalinity, and available inorganic nutrients like nitrogen.

Almost all soil organisms need food, water, and oxygen to live. Carbon-based food sources provide all the nutrients, including nitrogen and phosphorus, in a moist habitat with access to oxygen from the air. Unfortunately, some of the agricultural practices including land clearances, burning of the stubbles, inappropriate fertilizer applications, and over-tillage, have degraded the soils and produced conditions such as salinity, acidification, soil structural decline, and desertification.

SOIL REMEDIATION

Until recently, changing the pattern of farming practices was considered the only way to improve soil fertility. Recently, however, research studies have succeeded in inoculating soils and seeds with beneficial mycorrhizal fungi to improve their yields and to promote healthier soils. Although it is still at an early stage of development, the positive field trials are expected to improve soil fertility on a wider scale.

The functioning of the soil microorganisms, posing no threat to human health or agricultural production, was not examined in much detail, as soil microbiology in the past has mostly been concentrated on handling the harmful or pathogenic problems posed by soil microorganisms. Interdisciplinary soil research in the future is expected to include the interaction processes in which the soil is considered a living partner. Deploying non-destructive imaging, next-generation chemical analyses with substantial space and time resolution, and simulation modeling, the details of the soil and biological relationships would provide a holistic soil-science platform for understanding the plant-soil interactions to control agricultural sustainability and environmental changes favourably.

THE IMPORTANCE OF SOIL MICROORGANISMS

The changes in the structures and compositions of the microbial communities in the soil are considered indicators of soil biochemical processes and ecosystem functions. Soil nutrient cycles and turnover are determined by microbial communities, and any changes in these communities can affect microbial functions as well as soil nitrogen (N) and carbon (C) dynamics [6 - 9]. Soil enzymes play an equally important role in decomposing organic matter, transforming, and cycling the nutrients [10, 11]. Enzymes like hydrolases and oxidases are found decomposing substrates and releasing nutrients into the soil. The urease enzyme is associated with microbial N acquisition and it catalyzes the decomposition of urea. ß-1,4-glucosidase is a hydrolytic enzyme produced by soil microbes to decompose polysaccharides. Acidic and alkaline phosphatases associated with phosphorus (P) acquisition, cleave PO_4^{3-} from P-containing organic compounds. The enzyme arylsulfatase hydrolyses the organic sulphate esters, releasing SO_2 that is utilized by plants [6].

Microbial communities and soil enzymatic activities are found to be changed by the application of NMs and landscape positions [12]. Inorganic fertilizers are found to improve crop yields but also affect directly or indirectly the physical, chemical, and biological properties of the soil, as noted in varying soil microbial communities and enzymatic activities in agricultural and unmanaged ecosystems [13]. The addition of NMs to agricultural fields in one or a few large applications per year exceeds the rate of atmospheric deposition. Furthermore, the increase in crop productivity due to fertilization activity in agricultural systems increases inputs of organic material in the form of root exudates, decaying roots, and plant residues and thus increases the pool of C sources for soil microorganisms' growth [6]. Several field and laboratory studies have also identified major changes in the structure of the soil microbial community and soil enzymatic activities under N addition treatments [14].

Landscape positions have been found to influence soil properties [15]. Foot slope positions are generally rich in moisture, organic matter, and clay content, and thus tend to have higher primary productivity in comparison with upland areas. Footslope sites with more substrate, nutrients, and a suitable environment can support larger microbial communities with higher microbial biomass, favouring different microbial communities than upland sites and resulting in differences in the corresponding microbial communities. Soil enzymatic activities were found more productive with active microbial communities in foot slopes marked by the highest levels of soil moisture, plant biomass, N and C content compared to midslope positions. There is limited information available on the impact of NM-application rates and landscape positions on soil microbial communities and their enzyme activities in switchgrass. Understanding the effect of NMs on soil microbial communities and soil enzymatic activities is critically important for improving bioenergy crop production as soil microbiological properties are an integral component of soil health and productivity.

Bacteria and fungi function as chemoorganotrophs, obtaining carbon and energy by decomposing the organic compounds added to the soil, including plant residues and dead soil organisms. Those microorganisms secrete exo-enzymes to initiate the breakdown of litter consisting of compounds that are too large and insoluble to pass through microbial membranes. These exo-enzymes convert macromolecules into soluble products that are absorbed and metabolized by the microbes. Fungal biomass and activity are dominated by the microbial breakdown process of leaves in streams, while bacteria become relevant in advanced stages of the decomposition process. When microbes die, their bodies become part of the organic substrate available for decomposition. Bacteria and fungi are involved in the aerobic and anaerobic degradation of organic matter, but fungi are more efficient than bacteria in the degradation of highly recalcitrant organic matter because they produce a wider range of extracellular enzymes than bacteria. Laboratory studies have confirmed more rapid respirations in the upper layer [16].

NANOPARTICULATE MATERIALS-BRIEF INTRODUCTION

Particulate material species with at least one dimension of <100 nm are named nanoparticles (NPs). Nanoparticulate materials (NMs) are also defined in terms of their overall shapes, representing 0D, 1D, 2D, or 3D structures. These nanoparticulate materials are found to exhibit size-specific physicochemical properties as observed in 20-nm gold (Au), platinum (Pt), silver (Ag), and palladium (Pd) NPs, exhibiting characteristic wine red, yellowish gray, black, and dark black colours, respectively. NPs are in general composed of three layers, *viz.* (a) the surface layer, which is functionalized with a variety of smaller molecules, metal ions, surfactants, and polymers, (b) the shell layer, which is a chemically different material from the core in all aspects, and (c) the core, which is the central portion of the NP [17].

Classification of NPs

NPs are broadly divided into different categories according to their morphology, size, and chemical properties. Based on physical and chemical characteristics, some of the well-known classes of NPs include carbon-based NPs, metal-NPs, ceramic-NPs, semiconductors, and solid lipid-NPs.

Fullerenes and carbon nanotubes (CNTs) are two major classes of carbon-based NPs. Fullerenes have carbon atoms arranged into pentagonal and hexagonal bonded units, with each carbon atom sharing a sp^2-hybridized electron orbital. CNTs are tubular structures 1-2 nm in diameter that exhibit metallic or semiconducting properties. These structures are considered as seamlessly rolled graphite sheets with single, double, or many walls, and therefore, they are named single-walled (SWCNTs), double-walled (DWCNTs), or multi-walled carbon nanotubes (MWCNTs), respectively.

Metal NPs, made out of the metal precursors, possess localized surface plasmon resonance (LSPR) characteristics, exhibiting unique optoelectronic properties. The Cu, Ag, and Au-NPs have a broad absorption band in the visible region of the electromagnetic spectrum.

Ceramic NPs are inorganic, non-metallic solids in amorphous, polycrystalline, dense, porous, or hollow forms. These NPs are found to be very useful in catalysis, photocatalysis, photodegradation of dyes, and imaging applications [18].

Semiconductor NPs exhibiting size-dependent band gaps are used in photocatalysis, photo-optic, and electronic devices. As an example, a variety of semiconductor NPs are found exceptionally efficient in water splitting applications, due to their suitable bandgap and band edge positions.

After surface functionalization, nanospherical or nanocapsular organic NPs, also known as polymeric nanoparticles (PNPs), find a variety of applications, as discussed briefly below [19 - 21].

A lipid NP is spherical (*i.e.*, 10 to 1000 nm), with a solid core of lipid and a matrix containing soluble lipophilic molecules, stabilized by surfactants or emulsifiers. Lipid nanotechnology has been developed for the designing and synthesis of lipid-NPs for various applications, such as targeted drug delivery carriers releasing RNAs in cancer therapy.

Characterization of NPs

Advanced characterization techniques have been developed and used in analyzing various physicochemical properties of NPs by deploying X-ray diffraction (XRD), X-ray photoelectron spectroscopy (XPS), infrared (IR), Scanning electron microscopy (SEM), Transmission electron microscopy (TEM), Brunauer–Emmett–Teller (BET), and a particle size analyzer.

Morphological Characteristics

Of the different characterization techniques available for morphological studies, those involving polarized optical microscopy (POM), SEM, and TEM are the most commonly used. SEM has been used not only for the morphology of nanomaterials but also for their dispersion in the bulk or matrix. The dispersion of SWNTs in the polymer matrix of polybutylene terephthalate (PBT) and nylon-6 was measured using this technique [22]. Similarly, TEM is based on the electron transmittance principle, providing information about the bulk material from very low to high magnification. TEM has extensively been used in determining the morphologies of gold NPs.

Structural Characteristics

Structural properties of the NPs have been determined using XRD regarding the crystallinity and phase, besides giving a rough estimate of particle size [23 - 25]. This technique works well in both single-phase and multiphase NPs identification. Nevertheless, in the case of smaller NPs having a size of less than hundreds of atoms, the acquisition and correct measurement of structural and other parameters is rather difficult. Moreover, NPs having more amorphous characteristics with varying inter-atomic lengths have been found to influence the XRD measurement. In that case, proper comparison of the diffractograms of bimetallic NPs with those of the corresponding monometallic NPs and their physical mixtures helps in having accurate information. Comparison of computer-simulated structural models of bimetallic NPs with observed XRD spectra is the best way to get good contrast. EDX fixed with field emission scanning electron miscopy (FE-SEM) or TEM is used to know about the elemental composition with a rough idea of % wt. The electron beam focused on a single NP by SEM or TEM generates the intrinsic information of the NP. The constituent elements in NP emit characteristic energy

X-rays by electron beam irradiation with an intensity directly proportional to the concentration of the explicit element in the particle. This technique is widely used to give support to SEM and other techniques for confirming their elemental compositions [26, 27]. Using a similar technique, the elemental composition and graphene impregnation of the In_2O_3/graphene heterostructure was carried out, showing C, In, and O as contributing elements [19].

XPS, being capable of determining the elemental ratio and bonding nature of the elements in NP-material, has been deployed in depth profiling to know the overall composition and the compositional variation with depth. The XPS spectrum is a plot of the number of electrons *versus* the binding energy (eV). Each element has its own fingerprint binding energy value and thus gives a specific set of XPS peaks. Electronic interactions between metal-NPs and oxide support control the stability, activity, and selectivity of the catalysts involving electron transfer across the metal/support interface. In one such study, the charge transfer in a well-defined platinum/ceria catalyst was examined for heterogeneous catalysis. Combining several analytical tools like synchrotron-radiation photoelectron spectroscopy, STEM and DFT calculations, the charge transfer per Pt atom was the largest for Pt particles of around 50 atoms, *i.e.*, one electron per ten Pt atoms from the NP to the support. For smaller particles, charge transfer was found to be partially suppressed by defects [28].

Citrate stabilized Au/Ag-core or shell-NPs have been characterized using XPS and STEM to determine the composition and morphology. Core-shell type polydisperse Au/Ag-NPs of non-spherical shapes and containing off-cantered Au-cores were initially characterized using STEM analysis followed by SESSA spectral modeling showing that these NPs were stabilized with a 0.8 nm layer of sodium citrate and a 0.05 nm (one wash) or 0.025 nm (two wash) layer of hydrocarbon, but this did not fully account for the observed XPS signal from the Au-core. Many simulations were carried out to take into account the contributions of non-sphericity and off-centered Au-cores. The off-cantered cores were found to be critical and, after taking into consideration the geometrical nonuniformities in the simulations, the SESSA generated elemental compositions were found to match the XPS elemental compositions [29].

Vibrational characterization of NPs has been studied using FTIR and Raman spectroscopy. In one study, the functionalization of Pt-NPs (1.7 nm mean size) and their interaction with an alumina substrate were studied using FTIR and XPS techniques. FTIR confirmed the carboxylated functionalization by showing the vibrational peaks of Pr, La, Gd, Lu, and Er-NPs synthesized by correlated techniques. Using a semiconductor-NPs based substrate in SERS, it could be made to excite SPR in the visible region with a specific excitation laser source that makes the measurement easier and more reproducible than metal-NP-solutions. Semiconductors, being biocompatible could extend their usefulness for *in vivo* applications. The SERS technique was reported to examine the phonon modes in TiO_2, ZnO-NPs, and PbS-QDs, and the enhanced spectra were attributed to the plasmonic resonances in semiconductor systems [30].

Particle Size and Surface Area

Of the different techniques available for characterization of the NPs, those of SEM, TEM, XRD, and AFM provide information about the particle size, whereas the zeta potential size analyzer/DLS is used at extremely low levels [31]. The DLS technique was used to determine the size variation of silica NPs with the absorption of proteins from serum, showing increased size with the acquisition of protein layers. However, because DLS is inaccurate in the presence of agglomeration and hydrophilicity, it is better to use the high-resolution differential centrifugal sedimentation (DCS) technique. NP-tracking analysis (NTA) is another technique found helpful in biological systems of proteins and DNA by correlating the Brownian motion rate to particle size ranging from 10 to 1000 nm in a liquid medium [32].

Differential centrifugation is frequently used in the isolation and purification of viruses, cells, subcellular organelles, proteins, and nucleic acids or dispersed in relevant solvents, and was extended to NPs in successive steps with increasing centrifugation forces and durations to isolate the smaller particles from the larger ones. Larger particles were found to be removed in the first step of fast sedimentation, leaving most of the smaller particles in the supernatant. The supernatant applications of centrifugation steps as mentioned above were used in separating the rest to provide the sedimentation coefficients of the remaining particles were significantly different. It was found to be difficult to isolate particles with small differences in their sedimentation rates [33].

A number of methods for measuring the size, crystal structure, elemental compositions, and other physical properties of NPs are available. However, it is quite often necessary to use more than one method to take care of the strengths and shortcomings of each technique for having a combinatorial approach, as discussed in a comparative study reported recently [32].

The surface area of NPs was reported using nitrogen gas-producing four types of isotherms of adsorption and desorption in the Brunauer-Emmett-Teller (BET) method. By employing focused ion beam nano-slice tomography along with SEM imaging, NP-parameters like porosity, nanopore tortuosity, and electrical connectivity were measured at the single NP level. 3D-reconstruction of slices could reveal the pores with an average size of 3±2 nm and a relative nanopore tortuosity of 46.8±24.5. Electrical connectivity at the single-NP level was estimated because of material-selective electrodeposition as noted in Cu deposition on Pt-NPs adsorbed onto a carbon support. NP connectivity was determined by energy-dispersive X-ray spectroscopy and SEM. NPs synthesized by electrodeposition were found to show >97% connectivity with underlying highly oriented pyrolytic graphite (HOPG) or amorphous carbon electrodes [34].

Using colloidal silica and polystyrene latex as reference materials (20nm to 200nm), the validation of particle tracking analysis (PTA) was carried out in terms of the working range, linearity, the limit of detection and quantification, sensitivity, robustness, and precision. For nominal 50nm and 100nm polystyrene NPs and 80nm silica-NPs, the relative expanded measurement uncertainties were found to vary from 10% to 12%. In the presence of agglomerates in nominal 50nm polystyrene NPs, the uncertainty in the arithmetic mean of the particle size distributions was found to be up to 18% [31].

Optical Characterizations

Information about the absorption, reflectance, luminescence, and phosphorescence properties of metallic and semiconductor NPs using ultraviolet-visible (UV-Vis) spectrophotometry, photoluminescence (PL), and spectroscopic ellipsometry has been effective in exploring their optical properties. However, the UV/Vis-diffuse reflectance spectrometer (DRS) has successfully been used to measure the bandgaps of the NPs and other nanomaterials. Absorption shifts observed experimentally were thus used to determine doping, composite formation, and heterostructure NPs materials [35]. For instance, absorption characteristics obtained from the UV/Vis DRS technique in MMT, $LaFeO_3$, and $LaFeO_3$/MMT-nanocomposites showed a strong red shift compared to pristine MMT and $LaFeO_3$ NPs. $LaFeO_3$ and $LaFeO_3$/MMT displayed broad absorption bands from 400 to 620 nm with bandgap reduction. The $LaFeO_3$-NPs (10-15 nm) were deposited onto the surface of $LaFeO_3$/montmorillonite nanocomposites ($LaFeO_3$/MMT), causing efficient adsorption and excellent photocatalytic activity, with the overall removal rate of RhB up to 99.34% under visible light irradiation for 90 min. The study of interface characteristics and degradation mechanisms was explored, suggesting that $LaFeO_3$-NPs were immobilized on the surface of montmorillonite with the Si-O-Fe bonds, resulting in enhanced photocatalytic degradation [36].

Photocatalytic ZnO-NPs with wide bandgap and fast recombination rates of electron-hole pairs when doped with Er/Nd exhibited a suppressed recombination rate and extended absorption edge in the visible region with enhanced photocatalytic efficiency. The photocatalytic properties of the pure and doped

ZnO-NPs confirmed the photocatalytic activity of doped ZnO-NPs to be higher than the bare ZnO-NPs, possibly due to reduced recombination rate and efficient charge separation as inferred from the PL studies. It was further demonstrated that Nd-doped ZnO-NPs showed better antibacterial and anticancer efficacies as compared to pure and Er-doped ZnO-NPs [37].

Comparative PL-emissions from ZnO-NP and ZnO-Cu-NPs of different sizes showed PL-spectra corresponding to the near band edge emission accompanied by three and four defect-related PL bands in ZnO and ZnO-Cu-NPs, respectively. Their integrated PL intensities and the activation energies of PL thermal decays were estimated, showing that small Cu concentrations in ZnO-Cu-NPs stimulated the defect related PL bands [38].

Self-assembled hollow Au-NPs have been measured for refractive index and extinction coefficient and compared with those of solid Au-NPs, they exhibited strong SPR peaks located at wavelengths following a similar trend to those of the SPR positions of the different hollow Au-NPs in solution. The refractive index curves exhibited abnormal dispersion due to strong SPR extinction. The values of Δn and k_{max} were found to be linearly correlated with the particle number densities. Sensor behaviour was demonstrated by comparing the optical constant values of hollow Au-NPs with those of solid Au-NPs.

Physico-chemical Properties

Various physico-chemical properties, including large surface area to volume ratios, mechanical robustness, optically active and chemically reactive surfaces of the NPs, make them novel in various applications. Some of their important properties with respect to their soil interactions are discussed in brief as follows.

Electronic and Optical Properties

Noble metal-NPs having size-dependent optical properties were found to exhibit a strong UV-Visible extinction band that is not present in the spectrum of bulk metal. When the incident photon frequency coincides with the collective excitation of the conduction electrons, it gives rise to localized surface plasma resonance (LSPR). It is well established that the peak wavelength of the LSPR spectrum is dependent upon the size, shape, and interparticle spacing of the NPs as well as their dielectric properties and those of their local environment, including the substrate, solvents, and adsorbates. The free electrons on the surface of these NPs (d-electrons in Ag and gold) are transported freely through the nanomaterial with a mean free path of ~50 nm, which is more than the NP size. The absence of scattering from the bulk during light interaction causes a standing resonance condition, resulting in LSPR.

Magnetic Properties

Magnetic NPs have been examined for their applications in heterogeneous and homogeneous catalysis, biomedicine, magnetic fluids, data storage, magnetic resonance imaging (MRI), and environmental remediation like water decontamination [39, 40]. The uneven electronic distribution in NPs leads to magnetic properties dependent on the synthetic protocol and various synthetic methods used, such as solvothermal co-precipitation, micro-emulsion, thermal decomposition, and flame spray synthesis for their preparation [41].

Mechanical Properties

The mechanical parameters of NPs, including elastic modulus, hardness, stress, strain, adhesion, and friction, have been examined to know their mechanical nature in addition to their uses in surface coating, coagulation, and lubrication. NPs show different mechanical properties compared to their microparticle and bulk counterparts. Furthermore, in a lubricated or greased contact, the stiffness difference between the NPs and the contacting external surface is found to control whether the NPs are indented into the surface or deformed when high pressures are applied. A deep insight is, however, necessary to know the basics of the mechanical properties of NPs, such as elastic modulus, hardness, movement, friction, interfacial adhesion, and their size-dependent characteristics for their industrial applications.

Thermal Properties

The fluids containing suspended NPs were found to exhibit enhanced thermal conductivities relative to those of conventional heat transfer fluids. Nanofluids are produced by dispersing the NPs into liquids such as water, ethylene glycol, or oils. Because the heat transfer takes place at the surface of the NPs, it is desirable to use particles with a large total surface area that increases the stability of the suspension. Recently, it has been demonstrated that the nanofluids of CuO or Al_2O_3 NPs in water or ethylene exhibit advanced thermal conductivity.

NPS AND SOIL MICROBES

The presence of significant ambiguity in experimental observations reported by different research groups was bound to exhibit all kinds of mixed behaviour due to soil chemistry being highly complex and varying from location to location. The efforts to be made in this context would require tracing the interaction pathways followed by NP-species that are fairly complex and involved as compared to those involving other living organisms, including human beings. The rudimentary knowledge gained from simulated model experiments has recently begun to pick up due to the need to improve crop production in order to feed the world's growing population. Despite these odds, hybrid families of engineered nanomaterials involving inorganic-inorganic, inorganic-organic, inorganic-biopolymeric, and organic-biopolymeric NP-species combined with targeted delivery features have already begun to show some promise in providing cost-effective solutions in controlling soil microbial activities with site-specific applications. However, for all such attempts, it is necessary to delineate the interaction pathways of the NPs in the soil matrix with rich biodiversity. The only solution, which seems feasible at this stage, is to conduct a number of simple model experiments with designed soil samples before going in for large throughput analysis using in silico, artificial intelligence, or some other route of extracting the right kind of options for offering soil remediation solutions in the field [42].

To highlight the kinds of ambiguities reported recently in the context of NP interactions with soil, some observations made by the researchers are summarized here based on the recent observations by Usman and coworkers [5].

The interaction of engineered NMs (ENMs) with soil organic material (SOM) has been found to vary with their hydrophilic or hydrophobic nature and different biochemical compositions of the soil involved [43]. Contradictory responses of ENMs to SOM were reported primarily due to varying experimental conditions, soil properties, and ENM dosage in the experiments [12, 44, 45]. An insignificant impact on SOM was reported for low-dose exposure of Ag-NPs [45]. Similarly, metal-oxide NPs like CuO and Fe_3O_4 did not affect total SOM, except for changes in the biochemical composition, in contrast to TiO_2-NPs that did stabilize SOM by photo-oxidative coupling with humic molecules *via* covalent bonds [46]. Without influencing the microbe-derived SOM, TiO_2-NPs could reduce the decomposition because of the stability due to their complexation with HS [47]. The application of ZnO-NPs was found to reduce litter-derived organic carbon decomposition efficiency due to decreased microbial activity [48]. In another report, Fe_2O_3-NPs were found to suppress CO_2 emissions by up to 30%, indicating less SOM decomposition. A higher concentration of CuO-NPs could even decrease the SOM content in paddy soils [12].

Most of the microbial communities (*i.e.*, around 40-70% of total soil bacteria populations) have been found to reside in soil micro-aggregates available for interactions with the ENMs either due to their toxicity or by amplifying the bioavailability of other toxic compounds already present in the soil. The toxicity of these NPs against soil microbes has been reported depending on the type of NP, *e.g.*, metal and metal-oxide NPs exhibit higher toxicity compared to organic NPs like fullerenes and CNTs [47, 49]. In general, ENMs have been noted to influence

Gram-positive bacteria more than Gram-negative ones because of the different cell wall structures [50]. However, while Ag-NPs were found to be more toxic to Gram-negative bacteria, carbon-based NPs had a significant impact on the functional genes and pathways of soil microbes such as archaea, bacteria, and eukarya [51]. Similarly, metals and metal-oxide NPs acted as antimicrobial agents, inhibiting the growth of rhizobacteria and other bacterial species. TiO_2 and ZnO-NPs have been found to reduce the microbial biomass carbon (MBC), especially in Gram-negative bacteria [48, 49]. Soils exposed to different doses of TiO_2 and ZnO-NPs for over 60 days showed negative influences on substrate-induced respiration, leading to reduced microbial activity. Similarly, ZnO and Fe_2O_3 NPs were found to reduce MBC and the heterotrophic bacterial and fungal colonies that form units [48, 52]. CeO_2, Fe_3O_4, and SnO_2-NPs have no impact on MBC but microbial biodiversity was significantly changed. In contrast, Fe-oxide NPs stimulated the growth of actinobacteria by enhancing urease and invertase enzyme activities. A recent study also found that using the allowed concentration of ZnO and CuO-NPs (10 mg/kg) had a positive effect on microbes and enzyme activity [53]. The lower concentrations of CuO-NPs (10 and 100 mg/kg) did not show any inhibitory effect except denitrification [54].

These mixed influences of NPs on microbial biomass and their activities were found to be dependent on their doses, as it was shown that Ag-NPs had only a negative impact on microbial functioning with a high dose of Ag [45, 55]. Most of the Ag, Al_2O_3, TiO_2, CuO, and ZnO-NPs were found to have negative impacts on soil microbes, whereas those of Si, Fe, Au, Pd, and Ag_2S-NPs showed either no or very little influence [48, 50]. The toxicity of NPs to soil microbes was discovered to be dependent on particle size, dose, concentration, and nature of NPs, as well as soil type and moisture content [56, 57]. The NMs exceeding threshold concentrations were found to inhibit the growth of many of the soil microbes, affecting microbial biomass and their community structure [57, 58]. Deploying biogenic NMs was found to be less harmful to soil microbes compared to their chemically synthesized analogues, and thus, their use was promoted to address the nanotoxicity in soils [59, 60]. Due to the few studies available, more studies were recommended to explore further applications of biogenic NMs in soils [5].

Application of NPs at pre-optimized rates was found to improve seed germination, growth, and yield in several plant species. NPs also impart tolerance against different biotic and abiotic stresses in plants due to the expression of stress-tolerant genes and stress proteins [61].

The NPs penetrate the cell membrane and cell wall of the root epidermis through a complex series of events to enter the plant xylem, stele, and finally the leaves [62]. However, while passing through the intact cell membrane, NPs must move through size-specific pores. Before reaching the stele, NPs enter passively through the apoplast of the endodermis. The NP-uptake mechanism is mostly through active transport, including cellular processes such as signalling, recycling, and regulation of the plasma membrane [62]. NPs enter the plant root through osmotic pressure, capillary forces, and pores in the cell wall through plasmodesmatal connections and/or symplastic routes. Entries of the NPs in the plant cells are assisted by their binding to the carrier proteins, through ion channels, aquaporin, and endocytosis *via* the formation of new pores. Once the NPs enter the plant cell, they are transported *via* apoplast or symplast from one cell to another through plasmodesmata. The entry of NPs through the cell wall depends upon the pore size of the cell wall, and smaller sized NPs pass through them easily, while larger sized NPs penetrate through stomata, hydathodes, and flower stigmas [63]. The NPs are transported through stomata pores when the particle size is $\geq$40 nm and accumulate in the stomata instead of the vascular bundle before getting translocated to different parts *via* the phloem. The NPs enter through parenchymatous intercellular spaces in the seed coats. However, in the seed coat, the aquaporins are involved in regulating the NPs' entry [62].

NMS AND PLANTS

It has been observed that the type and concentration of NPs exposed to the plants cause several morphological changes in the plants with either positive or negative influences [64]. Enhanced nitrate reductase activity in TiO_2 and SiO_2-NP treated soybean seeds due to stimulated seed germination is a supporting example in this context. Similarly, TiO_2-NPs treated spinach exhibiting improved nitrogen assimilation and enhanced photosynthesis is another proof of concept. Soaking the pre-germinated wheat seeds in multiwall carbon nanotubes (MWCNTs) solution (100 $\pm$ 60 μg/l) for 4 hours exhibited faster root growth with larger biomass. Another study using Indian mustard seeds discovered that treatment in oxidized MWCNT solution (24.15$\pm$21.8μg/l) for 4 hours improved uniformity and rate of germination, resulting in root and shoot growth. Likewise, soaking watermelon seeds in Fe_2O_3-NPs could not only increase germination but also trigger plant growth and fruiting behaviour. SiO_2-NPs treated tomato seeds exhibited improved germination.

ENM-species find wider applications in conducting studies using *in vitro* cultures. For instance, ZnO-NPs showed a pronounced effect on the growth of tobacco (*Nicotiana tabacum* L.) calli. In addition to promoting calli growth, it also enhanced the protein content [65]. The foliar application of ZnO-NPs (10 mg/l) could increase the chlorophyll, total soluble leaf protein, and phosphorus concentration in cluster bean. Improving plant growth through faster germination and improved nitrogen-fixing ability of NP-species are currently being investigated for their expanding applications. Foliar application of SiO_2-NPs (5-15 nm; 300 µg/l) was reported to improve sugarcane growth by maintaining the effective quantum yield of cyclic electron flow during photosynthesis and photoprotection [66].

There are several reports on the beneficial effects of NPs, including some on phytotoxicity [65]. However, the beneficial effects and phytotoxicity aspects depend on dose, size and nature of NPs, duration and conditions of exposure [65, 67]. For instance, 25 nm sized CuO-NPs were more phytotoxic to soybeans than the larger sizes (50 and 250 nm) of CuO-NPs [68]. The NPs primarily affect seed germination, biomass, leaf number, and root elongation. NPs have also been shown to reduce seed germination, plant elongation and even plant death [69]. The negative effects reported include slow-growth, changed sub-cellular metabolism, oxidative damage to biological membranes [67], decreased photosynthesis [70], chromosomal abnormalities, disturbances in water transport, and water status of the plant [71], by adversely affecting plant growth hormones and changes in gene transcription profiles [72]. Plant cells are found to be influenced by the NPs by introducing changes in plant gene expression and associated biochemical pathways, which subsequently affect plant growth and development. Using TiO_2-NPs, for instance, caused damage to the genomic DNA. Transcriptomic analyses have revealed disturbances in the link for up and down regulation of genes in higher plants [73]. In maize, the exposure of SWCNT supregulated the SLR1 and RTCS genes, while down-regulated the RTH1 and RTH3 genes. The use of fullerenes was found to disturb the pathways of energy and transport of electrons by repressing the transcription genes [73]. Applications of some of the NPs showed upregulation of many genes associated with stress and water channels [72]. The use of MWCNTs, for instance, could up-regulate the genes NNtPIP1, NtLRX1, and CycB that are responsible for water transport, the formation of the cell wall, and cell divisions, respectively. In general, the application of a higher concentration of NPs is found to be toxic. For example, the application of CeO_2-NPs (2000 and 4000 mg/l) damaged the structure of DNA in soybean.

Application of TiO_2-NPs (2 mM concentration) was found to damage the DNA in tobacco, while at a higher concentration of 10mM it damaged the maize DNA. These results indicate the necessity of pre-testing these formulations before applying them in the field.

The exposure of metal-NPs to the plants may cause oxidative damage resulting in the production of reactive oxygen species (ROS) and the activation of antioxidant defense mechanisms [74]. The antioxidant defense mechanisms include both enzymatic (such as ascorbate peroxidase, catalase, superoxide dismutase, guaiacol peroxidase, and glutathione reductase) and non-enzymatic antioxidants (such as glutathione, ascorbate, thiols, and phenolics) [74, 75]. The catalase and guaiacol peroxidase are found to be reactive oxygen species (ROS)-quenching. Superoxide dismutase catalyzes the dismutation of superoxide anion into hydrogen peroxide [74]. The use of TiO_2-NPs may cause photocytotoxicity due to ROS generation. However, these generated ROS radicals may act as signalling molecules to activate the plant antioxidant defense mechanism to detoxify the free radicals. In NP-induced ROS generation, the ascorbate peroxidase reduces H_2O_2 into H_2O molecules [74]. Plants are also reported to develop antioxidant potential against NP-induced oxidative stress. The antioxidant enzymes are activated by several Fe_2O_4, CeO_2 and Co_3O_4 NPs inducing catalase; Fe_3O_4, CeO_2, MnO_2, CuO and Au-NPs inducing guaiacol peroxidase; while CeO_2, Pt-NPs, and fullerenes induce superoxide dismutase [62]. In the case of exposing spinach to TiO_2-NPs, improved activities were noted in superoxide dismutase, catalase, ascorbate peroxidase, and guaiacol peroxidase. Lower concentrations of TiO_2-NPs (200 mg/ml) were noted to increase chlorophyll, peroxidize catalase, malondialdehyde (MDA) content and superoxide dismutase through the elimination of ROS. Further increases in the concentration (500 mg/ml) of TiO_2-NPs resulted in cell membrane disruptions.

CONCLUDING REMARKS

While going through various approaches developed for understanding the basic mechanisms of NP-soil-microorganism interaction pathways, it is noted that there are still a number of gaps that need more detailed investigations using the currently available tools of high throughput data, analytical techniques for processing the experimental results collected from the different kinds of *in vitro*, *in vivo*, genomic, metabolomic, and proteomic tests for quantifying the cellular interactions in a fairly complex background of biodiversity possessed by the soil. Once the data available from simple models established in the laboratory experiments is available, further efforts might be made to simulate more complex systems successively. However, it is anticipated that the ambiguities reflected in numerous currently available results will be minimized and that these results could further be put to use in considering the possibility of improving the quality of crops with minimum interventions from the outside to mitigate the global shortages of food for human beings and animal stocks.

It is, thus, possible to conclude that the experience gained in developing nanomedicines would be extremely valuable in using nanocarriers for targeted deliveries that would optimize the requirements of materials involved in preparing nanoparticulate materials for improving and optimally utilizing soil fertility in producing better quality crops with higher yields.

CONSENT FOR PUBLICATION

Not applicable.

CONFLICT OF INTEREST

The author declares no conflict of interest, financial or otherwise.

ACKNOWLEDGEMENTS

The authors are extremely thankful to the Vice Chancellor, JC Bose University of Science & Technology, YMCA, for providing a congenial environment for carrying out the R & D activities in multidisciplinary studies.

REFERENCES

[1]   USDA. Sustainable Agriculture. 2021. https://nifa.usda.gov/topic/sustainable-agriculture

[2]   Jacoby R, Peukert M, Succurro A, *et al*. The role of soil microorganisms in plant mineral nutrition-current knowledge and future directions. Front Plant Sci 2017; 8: 1617.
[http://dx.doi.org/10.3389/fpls.2017.01617] [PMID: 28974956]

[3]   Ditta A. How helpful is nanotechnology in agriculture? Adv Nat Sci: Nanosci Nanotechnol 2012; 3: 033002.
[http://dx.doi.org/10.1088/2043-6262/3/3/033002]

[4]   Ahmad S. Engineered nanomaterials for drug and gene deliveries a review. J Nanopharm Drug Deliv 2015; 3: 1-50.

[5]   Usman M, Farooq M, Wakeel A, *et al*. Nanotechnology in agriculture: Current status, challenges and future opportunities. Sci Total Environ 2020; 721: 137778.
[http://dx.doi.org/10.1016/j.scitotenv.2020.137778] [PMID: 32179352]

[6]   Sekaran U, McCoy C, Kumar S, *et al*. Soil microbial community structure and enzymatic activity responses to nitrogen management and landscape positions in switchgrass (*Panicum virgatum* L.). Glob Change Biol Bioenergy 2019; 11: 836-51.

[7]   Leff JW, Jones SE, Prober SM, *et al*. Consistent responses of soil microbial communities to elevated nutrient inputs in grasslands across the globe. Proc Natl Acad Sci USA 2015; 112: 10967-72.
[PMID: 26283343]

[8]   Männistö M, Ganzert L, Tiirola M, *et al*. Do shifts in life strategies explain microbial community responses to increasing nitrogen in tundra soil? Soil Biol Biochem 2016; 96: 216-28.

[9]   Zeng J, Liu X, Song L, *et al*. Nitrogen fertilization directly affects soil bacterial diversity and indirectly affects bacterial community composition. Soil Biol Biochem 2016; 92: 41-9.

[10]   Jesus EDC, Liang C, Quensen JF, *et al*. Influence of corn, switchgrass, and prairie cropping systems on soil microbial communities in the upper Midwest of the United States. Glob Change Biol Bioenergy 2016; 8: 481-94.

[11]   Sherene T. Role of soil enzymes in nutrient transformation: a review. Bio Bulletin 2017; 3: 109-31.

[12]   Shi J, Ye J, Fang H, *et al*. Effects of copper oxide nanoparticles on paddy soil properties and components. Nanomaterials (Basel) 2018; 8: 839.
[http://dx.doi.org/10.3390/nano8100839] [PMID: 30332772]

[13]   Lladó S, López-Mondéjar R, Baldrian P. Forest soil bacteria: diversity, involvement in ecosystem processes, and response to global change. Microbiol Mol Biol Rev 2017; 81: e00063-16.

[http://dx.doi.org/10.1128/MMBR.00063-16] [PMID: 28404790]

[14]   Pan F, Zhang W, Liang Y, *et al.* Increased associated effects of topography and litter and soil nutrients on soil enzyme activities and microbial biomass along vegetation successions in karst ecosystem, southwestern China. Environ Sci Pollut Res Int 2018; 25: 16979-90.
[PMID: 29627959]

[15]   Jackson-Gilbert MM, Moses TM, Rao KPC, *et al.* Soil fertility in relation to landscape position and land use/cover types: a case study of the lake Kivu pilot learning site. Adv Agric 2015; 2015: 752936.
[http://dx.doi.org/10.1155/2015/752936]

[16]   Batubara SF, Agus F, Rauf A, *et al.* Soil respiration and microbial population in tropical peat under oil palm plantation. IOP Conf Ser Earth Environ Sci 2019; 260: 012083.
[http://dx.doi.org/10.1088/1755-1315/260/1/012083]

[17]   Shin W-K, Cho J, Kannan AG, *et al.* Cross-linked composite gel polymer electrolyte using mesoporous methacrylate-functionalized SiO_2 nanoparticles for lithium-ion polymer batteries. Sci Rep 2016; 6: 26332.
[http://dx.doi.org/10.1038/srep26332] [PMID: 27189842]

[18]   Khan I, Saeed K, Khan I. Nanoparticles: properties, applications and toxicities. Arab J Chem 2019; 12: 908-31.

[19]   Mansha M, Qurashi A, Ullah N, *et al.* Synthesis of In_2O_3/graphene heterostructure and their hydrogen gas sensing properties. Ceram Int 2016; 42: 11490-5.

[20]   Abouelmagd SA, Meng F, Kim BK, *et al.* Tannic acid-mediated surface functionalization of polymeric nanoparticles. ACS Biomater Sci Eng 2016; 2: 2294-303.
[PMID: 28944286]

[21]   Abd Ellah NH, Abouelmagd SA. Surface functionalization of polymeric nanoparticles for tumor drug delivery: approaches and challenges. Expert Opin Drug Deliv 2017; 14: 201-14.
[PMID: 27426638]

[22]   Saeed K, Khan I. Characterization of clay filled poly (butylene terephthalate) nanocomposites prepared by solution blending. Polímeros 2015; 25: 591-5.

[23]   Khan I, Ali S, Mansha M, *et al.* Sonochemical assisted hydrothermal synthesis of pseudo-flower shaped Bismuth vanadate ($BiVO_4$) and their solar-driven water splitting application. Ultrason Sonochem 2017; 36: 386-92.
[PMID: 28069225]

[24]   Khan I, Ibrahim AAM, Sohail M, *et al.* Sonochemical assisted synthesis of RGO/ZnO nanowire arrays for photoelectrochemical water splitting. Ultrason Sonochem 2017; 37: 669-75.
[PMID: 28427681]

[25]   Ullah H, Khan I, Yamani ZH, *et al.* Sonochemical-driven ultrafast facile synthesis of SnO_2 nanoparticles: Growth mechanism structural electrical and hydrogen gas sensing properties. Ultrason Sonochem 2017; 34: 484-90.
[PMID: 27773272]

[26]   Avasare V, Zhang Z, Avasare D, *et al.* Room-temperature synthesis of TiO_2 nanospheres and their solar driven photoelectrochemical hydrogen production. Int J Energy Res 2015; 39: 1714-9.

[27]   Iqbal N, Khan I, Yamani ZH, *et al.* Sonochemical assisted solvothermal synthesis of gallium oxynitride nanosheets and their solar-driven photoelectrochemical water-splitting applications. Sci Rep 2016; 6: 32319.
[http://dx.doi.org/10.1038/srep32319] [PMID: 27561646]

[28]   Lykhach Y, Kozlov SM, Skála T, *et al.* Counting electrons on supported nanoparticles. Nat Mater 2016; 15: 284-8.
[PMID: 26657332]

[29] Wang Y-C, Engelhard MH, Baer DR, *et al.* Quantifying the impact of nanoparticle coatings and nonuniformities on XPS analysis: gold/silver core-shell nanoparticles. Anal Chem 2016; 88: 3917-25. [PMID: 26950247]

[30] Muehlethaler C, Leona M, Lombardi JR. Review of surface enhanced Raman scattering applications in forensic science. Anal Chem 2016; 88: 152-69. [PMID: 26638887]

[31] Kestens V, Bozatzidis V, De Temmerman P-J, *et al.* Validation of a particle tracking analysis method for the size determination of nano- and microparticles. J Nanopart Res 2017; 19: 271. [http://dx.doi.org/10.1007/s11051-017-3966-8] [PMID: 28824287]

[32] Mourdikoudis S, Pallares RM, Thanh NTK. Characterization techniques for nanoparticles: comparison and complementarity upon studying nanoparticle properties. Nanoscale 2018; 10: 12871-934. [http://dx.doi.org/10.1039/c8nr02278j] [PMID: 29926865]

[33] Livshits MA, Khomyakova E, Evtushenko EG, *et al.* Isolation of exosomes by differential centrifugation: Theoretical analysis of a commonly used protocol. Sci Rep 2015; 5: 17319. [http://dx.doi.org/10.1038/srep17319] [PMID: 26616523]

[34] Matthew L, Glasscott W, Pendergast AD, *et al.* Advanced characterization techniques for evaluating porosity, nanoporetortuosity, and electrical connectivity at the single nanoparticle. ACS Appl Nano Mater 2019; 2: 819-30.

[35] Makuła P, Pacia M, Macyk W. How to correctly determine the band gap energy of modified semiconductor photocatalysts based on UV–vis spectra. J Phys Chem Lett 2018; 9: 6814-7. [PMID: 30990726]

[36] Peng K, Fu L, Yang H, *et al.* Perovskite LaFeO$_3$/montmorillonite nanocomposites: synthesis, interface characteristics and enhanced photocatalytic activity. Sci Rep 2016; 6: 19723. [http://dx.doi.org/10.1038/srep19723] [PMID: 26778180]

[37] Raza W, Faisal SM, Owais M, *et al.* Facile fabrication of highly efficient modified ZnO photocatalyst with enhanced photocatalytic, antibacterial and anticancer activity. RSC Advances 2016; 6: 78335-50.

[38] Torchynska TV, El Filali B, Ballardo Rodríguez IC, *et al.* Defect related emission of ZnO and ZnO Cu nanocrystals prepared by electrochemical method. PSS 2016; 13: 594-7.

[39] Faivre D, Bennet M. Materials science: Magnetic nanoparticles line up. Nature 2016; 535: 235-6. [PMID: 27411627]

[40] Priyadarshana G, Kottegoda N, Senaratne A, *et al.* Synthesis of magnetite nanoparticles by top-down approach from a high purity ore. J Nanomater 2015; 2015: 317312. [http://dx.doi.org/10.1155/2015/317312]

[41] Qi M, Zhang K, Li S, *et al.* Superparamagnetic Fe$_3$O$_4$ nanoparticles: synthesis by a solvothermal process and functionalization for a magnetic targeted curcumin delivery system. New J Chem 2016; 40: 4480-91.

[42] Wei Z, Gu Y, Friman V-P, *et al.* Initial soil microbiome composition and functioning predetermine future plant health. Sci Adv 2019; 5: eaaw0759. [http://dx.doi.org/10.1126/sciadv.aaw0759] [PMID: 31579818]

[43] Grillo R, Rosa AH, Fraceto LF. Engineered nanoparticles and organic matter: a review of the state-o--the-art. Chemosphere 2015; 119: 608-19. [PMID: 25128893]

[44] Schlich K, Hund-Rinke K. Influence of soil properties on the effect of silver nanomaterials on microbial activity in five soils. Environ Pollut 2015; 196: 321-30. [PMID: 25463729]

[45] Rahmatpour S, Shirvani M, Mosaddeghi MR, *et al.* Dose–response effects of silver nanoparticles and silver nitrate on microbial and enzyme activities in calcareous soils. Geoderma 2017; 285: 313-22.

[46]   Nuzzo A, Madonna E, Mazzei P, *et al.* *In situ* photo-polymerization of soil organic matter by heterogeneous nano-TiO_2 and biomimetic metal-porphyrin catalysts. Biol Fertil Soils 2016; 52: 585-93.

[47]   Simonin M, Guyonnet JP, Martins JMF, *et al.* Influence of soil properties on the toxicity of TiO_2 nanoparticles on carbon mineralization and bacterial abundance. J Hazard Mater 2015; 283: 529-35. [PMID: 25464292]

[48]   Rashid MI, Shahzad T, Shahid M, *et al.* Zinc oxide nanoparticles affect carbon and nitrogen mineralization of *Phoenix dactylifera* leaf litter in a sandy soil. J Hazard Mater 2017; 324: 298-305. [PMID: 27810328]

[49]   Rajput V, Minkina T, Sushkova S, *et al.* ZnO and CuO nanoparticles: a threat to soil organisms, plants, and human health. Environ Geochem Health 2020; 42: 147-58. [PMID: 31111333]

[50]   McKee MS, Filser J. Impacts of metal-based engineered nanomaterials on soil communities. Environ Sci Nano 2016; 3: 506-33.

[51]   Wu F, You Y, Werner D, *et al.* Carbon nanomaterials affect carbon cycle-related functions of the soil microbial community and the coupling of nutrient cycles. J Hazard Mater 2020; 390: 122144. [http://dx.doi.org/10.1016/j.jhazmat.2020.122144] [PMID: 32006845]

[52]   Rashid MI, Shahzad T, Shahid M, *et al.* Toxicity of iron oxide nanoparticles to grass litter decomposition in a sandy soil. Sci Rep 2017; 7: 41965. [PMID: 28155886]

[53]   Jośko I, Oleszczuk P, Dobrzyoska J, *et al.* Long-term effect of ZnO and CuO nanoparticles on soil microbial community in different types of soil. Geoderma 2019; 352: 204-12.

[54]   Zhao S, Su X, Wang Y, *et al.* Copper oxide nanoparticles inhibited denitrifying enzymes and electron transport system activities to influence soil denitrification and N_2O emission. Chemosphere 2020; 245: 125394. [http://dx.doi.org/10.1016/j.chemosphere.2019.125394] [PMID: 31862554]

[55]   He S, Feng Y, Ni J, *et al.* Different responses of soil microbial metabolic activity to silver and iron oxide nanoparticles. Chemosphere 2016; 147: 195-202. [PMID: 26766356]

[56]   Chen M, Sun Y, Liang J, *et al.* Understanding the influence of carbon nanomaterials on microbial communities. Environ Int 2019; 126: 690-8. [PMID: 30875562]

[57]   Peng F-J, Pan C-G, Zhang N-S, *et al.* Benthic invertebrate and microbial biodiversity in sub-tropical urban rivers: Correlations with environmental variables and emerging chemicals. Sci Total Environ 2020; 709: 136281. [http://dx.doi.org/10.1016/j.scitotenv.2019.136281] [PMID: 31905563]

[58]   Kang S, Zhu Y, Chen M, *et al.* Can microbes feed on environmental carbon nanomaterials? Nano Today 2019; 25: 10-2.

[59]   Mishra S, Yang X, Singh HB. Evidence for positive response of soil bacterial community structure and functions to biosynthesized silver nanoparticles: An approach to conquer nanotoxicity? J Environ Manage 2020; 253: 109584. [http://dx.doi.org/10.1016/j.jenvman.2019.109584] [PMID: 31634747]

[60]   Ottoni CA, Lima Neto MC, Léo P, *et al.* Environmental impact of biogenic silver nanoparticles in soil and aquatic organisms. Chemosphere 2020; 239: 124698. [http://dx.doi.org/10.1016/j.chemosphere.2019.124698] [PMID: 31493753]

[61]   Van Aken B. Gene expression changes in plants and microorganisms exposed to nanomaterials. Curr Opin Biotechnol 2015; 33: 206-19.

[PMID: 25827116]

[62] Tripathi DK, Shweta , Singh S, *et al.* An overview on manufactured nanoparticles in plants: Uptake, translocation, accumulation and phytotoxicity. Plant Physiol Biochem 2017; 110: 2-12.
[PMID: 27601425]

[63] Hossain Z, Mustafa G, Sakata K, *et al.* Insights into the proteomic response of soybean towards Al$_2$O$_3$, ZnO, and Ag nanoparticles stress. J Hazard Mater 2016; 304: 291-305.
[PMID: 26561753]

[64] Siddiqui MH, Al-Whaibi MH, Firoz M, *et al.* Role of Nanoparticles in Plants.Nanotechnology and Plant Sciences. Siddiqui M, Al-Whaibi M, Mohammad F, Eds. Cham: Springer 2015; pp. 19-35.

[65] Mazaheri-Tirani M, Dayani S. *In vitro* effect of zinc oxide nanoparticles on *Nicotiana tabacum* callus compared to ZnO micro particles and zinc sulfate (ZnSO$_4$). Plant Cell Tissue Organ Cult 2020; 140: 279-89.

[66] Elsheery NI, Sunoj VSJ, Wen Y, *et al.* Foliar application of nanoparticles mitigates the chilling effect on photosynthesis and photoprotection in sugarcane. Plant Physiol Biochem 2020; 149: 50-60.
[PMID: 32035252]

[67] Noori A, Donnelly T, Colbert J, *et al.* Exposure of tomato (*Lycopersicon esculentum*) to silver nanoparticles and silver nitrate: physiological and molecular response. Int J Phytoremed 2020; 22: 40-51.
[PMID: 31282192]

[68] Yusefi-Tanha E, Fallah S, Rostamnejadi A, *et al.* Particle size and concentration dependent toxicity of copper oxide nanoparticles (CuONPs) on seed yield and antioxidant defense system in soil grown soybean (*Glycinemax* cv. Kowsar). Sci Total Environ 2020; 715: 136994.
[http://dx.doi.org/10.1016/j.scitotenv.2020.136994] [PMID: 32041054]

[69] Yang X, Pan H, Wang P, *et al.* Particle-specific toxicity and bioavailability of cerium oxide (CeO$_2$) nanoparticles to Arabidopsis thaliana. J Hazard Mater 2017; 322: 292-300.
[PMID: 27021431]

[70] Barhoumi L, Oukarroum A, Taher LB, *et al.* Effects of superparamagnetic iron oxide nanoparticles on photosynthesis and growth of the aquatic plant *Lemna gibba*. Arch Environ Contam Toxicol 2015; 68: 510-20.
[PMID: 25392153]

[71] Martínez-Fernández D, Barroso D, Komárek M. Root water transport of *Helianthus annuus* L. under iron oxide nanoparticle exposure. Environ Sci Pollut Res Int 2016; 23: 1732-41.
[PMID: 26396006]

[72] García-Sánchez S, Bernales I, Cristobal S. Early response to nanoparticles in the *Arabidopsis* transcriptome compromises plant defence and root-hair development through salicylic acid signalling. BMC Genomics 2015; 16: 341.
[http://dx.doi.org/10.1186/s12864-015-1530-4] [PMID: 25903678]

[73] Landa P, Prerostova S, Petrova S, *et al.* The transcriptomic response of *Arabidopsis thaliana* to zinc oxide: a comparison of the impact of nanoparticle, bulk, and ionic zinc. Environ Sci Technol 2015; 49: 14537-45.
[PMID: 26560974]

[74] Rico CM, Peralta-Videa JR, Gardea-Torresdey JL. Chemistry, Biochemistry of Nanoparticles, and Their Role in Antioxidant Defense System in Plants. Siddiqui M, Al-Whaibi M, Mohammad F, Eds. Nanotechnology and Plant Sciences. Cham: Springer 2015; pp. 1-17.

[75]   Singh J, Lee B-K. Influence of nano-TiO$_2$ particles on the bioaccumulation of Cd in soybean plants (*Glycine max*): A possible mechanism for the removal of Cd from the contaminated soil. J Environ Manage 2016; 170: 88-96.
[PMID: 26803259]

CHAPTER 15

Intellectual Property Rights, Regulatory Laws, WTO, and its Impact

Bineeta Devi[1], Ashutosh Kumar[2,*] and K.R. Maurya[3]

[1] *Department of Genetics and Plant Breeding, Faculty of Agriculture Science and Technology, AKS University, Satna, M.P., India*

[2] *Department of Agriculture Economics, Faculty of Agriculture Science and Technology, AKS University, Satna, M.P., India*

[3] *Department of Horticulture, Faculty of Agriculture Science and Technology, AKS University, Satna, M.P., India*

Abstract: This document examines major economic issues for intellectual property rights protection (IPRs) in the context of the World Trade Organization (WTO). The idea is to look back on the establishment of TRIPS (trade-related aspects of intellectual property rights), which is an ongoing success of the Uruguay Round debate on free trade. This paper reviews the economic concept of harmonizing IPRs, drawing attention to economic theory and strong emerging evidence. The concept of linking IPR protection with trade in the context of the WTO is also being explored. Particular attention has been paid to the results of TRIPS on new agricultural and biotechnology innovations. The impact of IPR protection on promoting growth and development and the relationship between IPRs and other economic policies are discussed. This paper concludes with an analysis of the potential for more (or less) consensus related to IPRs in the current WTO negotiating cycle.

Keywords: Intellectual property rights, Patients, Trade mark, TRIPS, WTO.

INTRODUCTION

The term "patent" is derived from the Latin word *"litterae patentes"*, which means "an open letter". The Crown in England used to grant a right to an individual by writing a document with the seal of the King or Queen. Such grants were known as 'Letters Patent'. This was a letter that was rolled up and not sealed. Nowadays, the word "patent" is used as a synonym for the monopoly right associated with an invention. In French, "Brevet" (Latin, meaning "brief letters") is the document that grants rights or privileges for an identified invention. Thus,

* **Corresponding author Ashutosh Kumar:** Department of Agriculture Economics, Faculty of Agriculture Science and Technology, AKS University, Satna, M.P., India; E-mail: kumar.ashu777@gmail.com

Ashutosh Gupta, Shampi Jain & Neeraj Verma (Eds.)

patents date back several years in different forms with legal rights and provide new inventions based on scientific and technological know-how. It does not permit the inventor to commercialize the invention and prevents others from using and benefitting from it.

The patent protects the first producer or inventor of an article against manufacturing, using, or selling the article without his/her consent. It should also be clear that a patent is not granted for such an idea or policy but can be granted for an article or a process for making an article through an opinion. A common misconception about the patent right is that it gives an inventor the absolute right to exploit the invention, which is not so. The exploitation of the right also depends on whether others have a patent that overlaps with the subject matter of the invention. It also depends on other existing laws, such as those concerning health and safety. The basic patent provisions under the terms of trade-related intellectual property rights (TRIPs) agreement provide that the copyright holder has a legal right to restrict others from making, using, or selling new inventions for a limited period, with several exceptions, which we will study in this chapter. It, however, does not provide information on whether the product is safe for the customer or not.

The invention can be seen throughout nature and is a well-recognized phenomenon that can be seen in birds and other animals in the construction of their dwelling places with intricacy. For example, the tailorbird constructs a well-weaved nest, bees construct honeycombs, and the spider makes its web, each of which is unique in its style. They all do it without any risk of appropriation of the invention of one species by another. Infringements are committed by man alone, and so an invention requires protection from unauthorized copying or commercial exploitation, which hampers the rights and profits of the real inventor.

The most important aspect of the design is that it should be useful, novel, and unobtrusive. It is an agreement between a developer and a government. In this case, the government grants limited copyright to the designer without excluding others from using, selling, or producing the invention. But it has the condition that the details of the invention have to be disclosed by the inventor in the application for filing the patent. Thus, the whole patent can be thought of as a contract between the inventor and the state. Both of them bring consideration to that contract.

Under this social contract, a person applying for the patent brings consideration in terms of fees and, finally, after getting the patent consideration as royalty. As a result, the inventor is given the exclusive rights to prevent others from using, selling, or manufacturing the patented invention/articles for a fixed period and, as a result, gets the reward for disclosing his invention to the public. Ultimately, it can be concluded that the patent is not only a patentee's consideration but also a consideration of the state as well as the country. It is not possible to separate the patent from the state.

TYPES OF INTELLECTUAL PROPERTY RIGHTS

Intellectual property (IP) can be called a gift to mankind. Through this, the creation of special natural gifts and innovations that benefit mankind can be designed. But it is also important to protect the exploitation of such an invention and protect the rights of the originator. As it is common knowledge that any property, movable or immovable, needs to be legally protected to prevent it from getting stolen, similarly, the rights in the intellectual property created need to be protected to prevent it from infringement. There can be various kinds of innovations based on research, ideas, thoughts, graphics, designs, the way of writing a logo, and many more.

The different types of intellectual property are shown in Fig. (**1**). Patents, designs, trademarks, copyright, and geographical indications follow their rights and terms of protection.

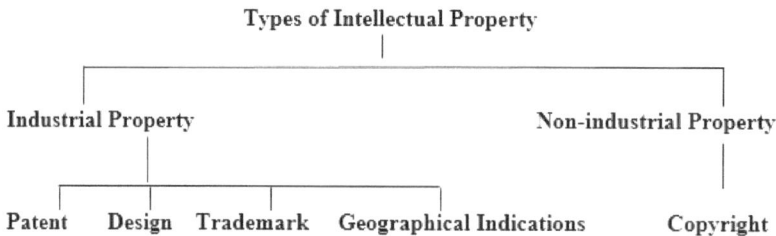

Types of Intellectual Property

Industrial Property	Non-industrial Property
Patent Design Trademark Geographical Indications	Copyright

Fig. (1). Types of intellectual properties.

Patents

A patent is the granting of a property right by a sovereign authority to an inventor. In exchange for a thorough disclosure of the invention, the inventor receives exclusive rights to the patented process, design, or invention for a set period of time. A patent is issued by the government to give the inventor exclusive rights for a limited period of time (20 years for a patent), for their innovation. The subject of a patent, which involves scientific and legal issues, is relatively complicated as compared with other types of intellectual property like designs, trademarks, and copyrights. It involves inventions that are new, useful, industrially applicable, and non-obvious, with an exhaustive list of non-patentable inventions. Some important examples of patents are calculators, televisions, elevators, radios, anaesthetic drugs, diesel engines, sewing machines, typewriters, motion pictures, air conditioners, computers, *etc.*

In India, at present, the 1970 patent law and related laws regulate the granting of patents. The Indian Parliament has made several amendments to the Copyright Act of 1970 in 1999, 2002, 2005, and 2006. The Patent Act 2003 was amended in 2005 and 2006, and a framework based on the size and number of claims, in addition to the basic fees, was introduced. The 2005 amendment has major implications for the introduction of patent protection of food, chemical, and agrochemicals and testing of "mailbox" applications. This was allowed under the TRIPs obligation to introduce product patenting in these sectors, latest by 1 January 2005. Until then, only process patents were allowed in India. As of date, India is fully compliant with its international obligations under the TRIPs agreement.

The Indian Patent Act of 1970 grants the following rights to the patentee:

• Rights to exploit the patent.
• Rights to license and assign the patent.
• Rights to surrender the patent.
• Rights to sue for the infringement of a patent.

The patent system does not protect every inventor who conceives an invention. The patent is granted based on the right to first disclosure of an invention. Whosoever tries to hinder the rights of the patentee suffers from the remedies available for the infringement. There are legal remedies or punishments available for the patentee, like interlocutory or interim injunction, damages or accounts of profits, or permanent injunction.

Prerequisites for Patent

A patent is an invention that is novel, useful, and industrially productive and which is so valuable that it requires protection from infringement. It is also important to know that not every invention is patentable. In the forthcoming chapters, non-patentable inventions, those inventions for which a patent can not be provided, are covered. For patentability, there are only three prerequisites according to the Section 2(1)(ac) of the Indian Patent Act, which are as follows:

1. An invention must be novel.
2. An invention should have an inventive step.
3. An invention should be capable of industrial application.

According to US patent law, Section 101 of the US Patent Act, inventions are patentable when they fulfill the criteria of:

- Novelty
- Usefulness and
- Non-obviousness

Novelty means that the inventions should be new or innovative, *i.e.* they should not be available to the public earlier. An invention must be an inventor's discovery. It can be either an improvement on existing articles or methods or even a small functional improvement. For example, a design patent protects the improved appearance of the product, while a utility patent protects the functional improvement to the existing product or a process. However, if the difference between the pre-existing and the new product or process is not sufficient, it will not get the patent. If the inventor or anyone else has publicly disclosed the claimed subject matter of the invention more than one year before the filing date of the patent application, then such an invention will be barred from receiving a patent.

Usefulness is another requirement for applying for patents. The invention should have some industrial applicability and provide benefits to the masses.

The non-obviousness of patents is considered if the inventor gets an unexpected outcome from the combination of known prior art elements with their known characteristics. For example, combining the molecules that comprise "minoxidil" resulted in maintaining good blood pressure, but if by using the same medicine, hair growth occurs, which was unexpected, it is a non-obvious and patentable combination of molecules. Non-obviousness and inventive steps are the two terms that reflect the same general patentability requirement present in most patent laws, according to which an invention, to be patentable, should be sufficiently inventive or non-obvious. European and Indian patent laws require an invention to involve an inventive step, while US patent law requires an invention to be non-obvious. Though this may seem at first to be essentially the same, there are important differences.

In Europe, the assessor determines the differences between the innovation and the prior art. If there is no difference found, the invention is not novel. However, if there is a difference, the examiner decides which technical problem is solved by adding these items to the previous art program. If no technical problem is found, it is considered that the invention does not involve an inventive step. While in the USA, the examiner checks the obviousness of the combination of the novel elements without looking for a solution to a technical problem.

Nature of Patent

Patents are of two different types. On the one hand, it acts as a piece of personal property and holds all the rights that can be applicable to a property, while on the other hand, it acts as a document that is concerned with the details of the invention and legal formalities.

Patent as a Form of Property

It is considered a piece of personal property and possesses all the characteristics of any other type of property (movable or immovable). Like any other business commodity, it may be bought, sold, mortgaged, or licensed. It can also be bequeathed or passed to the heirs of a deceased patentee. Therefore, anyone can make money out of his or her intellectual property by treating it as property. Thus, the patent right is a tradable commodity, which results in profiting the inventor commercially. Unlike any other property, the term of protection of such intellectual property is limited. The inventor loses all his or her rights to the property after the patent expires. The justification given is that the inventors are rewarded for their time, work, and risk of capital by the grant of a limited, though strong monopoly.

Patent as a Form of Document

A patent is a document issued by the statutory authority to the patentee that enables him/her to possess legal rights over his or her invention. The patentee is authorized to commercially exploit the invention. This document is the most important document submitted by the applicant. It is a technical as well as a legal document. The document must contain the description of the invention and claims, along with the drawings and figures required to explain the invention. The document consists of several forms, some of which are mandatory, while others are optional. These forms collect the information regarding the inventor's details, the details of the invention, power of attorney, statement, and undertaking under Section 8.

The following documents are required to be filed at the Indian patent office:

- From 26 (letter of authorization) on Indian Stamp paper duly signed by the inventor(s) assignee/applicant.
- Search report, if any.
- A preliminary certified copy of the priority documents.
- International search report, if any.
- Proof of rights.

The form contains the application number, date of application, the name and address of the applicant or inventors, agent and address for service, classification of the invention, field of research, title abstract, *etc.* The document specifies that the patent is protected from imitation by others. The document is composed of parts like

- Specification/description
- Claim and
- Grant

Specification

The specification and claim are published as a single document, which is available to the public at a minimal charge from the patent office. The description must disclose the invention sufficiently enough so that any person skilled in the particular branch of learning will be able to reproduce the same result. It should give a narrative description of the subject and should explain how the invention is carried out. The specifications are of the following two types:

a. Provisional: A provisional specification does not fully disclose the invention. It is taken to claim the priority date of an invention when the invention needs time to develop further.
b. Complete: In the complete specification, the document should contain a detailed description of the invention along with the drawings and claims. The description regarding prior art is also included.

The patent specification must be filled out in a specific format. The format contains the following headings:

- Title: It should mention the name of the invention, which denotes, in general terms, the technical field of the invention.
- Field of invention: It should clearly describe in a few sentences the broad area or field of the invention in which it falls.
- Description of current techniques: It should describe the current systems or methods used in relation to the inventions in brief.
- Problems with the current system or method: It should describe the shortcomings of the current system, technique, or methods like poor performance, inaccuracy of results, cost, problems in manufacturing due to the current process, *etc.*
- Proposed solution: It should explain its advantages over the current system. The main features of the invention's description must be given. It includes the

essential as well as preferred features. The essential features are those that solve the existing problems, whereas the preferred add-ons enhance the performance of the invention.

- Explain the solution in detail with at least one example: The inventor has to provide clear and full technical details of the invention at the time of filling out the application. The details of the invention must be included in the description because, after submission, the inventor is not allowed to add any technical information. The patent specification allows adding sketches, flow charts, and diagrams, which help in understanding the invention clearly.
- Alternative techniques or modifications: It should also explain the alternative techniques, modifications, or additions.
- Wider use or development of a solution: It should also mention the wider use or applications of the invention, which can help in solving the problem with enormous results.
- Outline of advantages: If the advantages were not mentioned under the headings above, their preferred features should be specified at the end of the description. It is equally important to describe the disadvantages of the inventions, if any.

Claim: A patent claim is defined as the extent of a monopoly right that an applicant holds. It is that part of the document that may not be practised by others. Inventors may claim a part or all of that which is described in the specification. The claim generally covers the following:

A product: The claim covers a product with all its uses and those that are yet to be discovered. For example, a novel drug that was patented for the cure of cancer was later found to cure heart disease; the patent will cover this new use also.

Use: The claim covers only the specific use of a product. Unlike the above-said drug, the claim will protect only its use to cure cancer and not its new use to cure heart disease. In some countries, new uses of existing inventions are patentable. If the patent for the existing invention is still alive, then the owner of the new use of the invention will have to acquire a license from the current patentee to exploit his new invention.

A process: The claim will protect the process of manufacturing the product but will not protect the product manufactured by that process.

A product of a process: The claim will protect only those products that are manufactured by the process described in the patent application. Therefore, it would cover the drug when made by a specified process.

Grant: The grant is filed at the patent office and is not published. It is a signed document and is the agreement that grants patent rights to the inventor.

After completing the documents, they are submitted either online at the website http://ww.inpindiaonline.gov.in/on_line/ or the true copies or hard copies can be submitted to the patent office. There are four patent offices for each territory where patents can be filed. Table **1** gives details of the addresses of the patent offices with their respective territorial jurisdiction.

Forms for a Patent Application

After the applicant identifies the patent office, he or she should file the patent application along with the requisite documents as discussed earlier. The following is an overview of some of the forms that have to be submitted.

Form 1– Application for Grant of Patent

As the name suggests, this form is an application for a grant of a patent in India. In this form, the applicant provides information such as the inventor's name and address, the applicant's name and address, information corresponding to prior patent applications relating to the current inventions filed by the applicant or any authorized entity, and some declarations, among other things.

Table 1. Patent offices in India.

Office	Address	Territorial Jurisdiction
New Delhi	Intellectual Property Office, Intellectual Property Office Building, Plot No. 32, Sector 14, Dwarka, New Delhi-110075. E-mail: delhi-patnt@nic.in	The States of Haryana, Himachal Pradesh, Jammu and Kashmir, Punjab, Rajasthan, Uttar Pradesh, Uttaranchal, Delhi, and the Union Territory of Chandigarh
Chennai	Intellectual Property Office, Intellectual Property Office Building, G.S.T. Road, Guindy, Chennai – 600032 E-mail: Chennai-patent@nic.in	The states of Andhra Pradesh, Karnataka, Kerala, Tamil Nadu, and the Union Territories of Pondicherry and Lakshadweep
Mumbai	Intellectual Property Office, Boudhik Sampada Bhawan, Near Antop Hill Post Office, S.M. Road, Antop Hill, Mumbai-400037. E-mail: Mumbai-patent@nic.in	The States of Maharashtra, Gujarat, Madhya Pradesh, Goa, and Chhattisgarh and the Union Territories of Daman and Diu and Dadra & Nagar Haveli
Kolkata	Intellectual Property Office, Intellectual Property Office Building, CP-2, Sector V, Salt Lake City, Kolkata-700091 E-mail: patent@nic.in	The rest of India.

(Source: WWW.ipindia.gov.in)

Form 2–Provisional/Complete Specification

This form furnishes the patent specification. The patent specification can be provisional or complete, depending on the type of patent application (provisional or complete). The specification particularly describes the invention and how it was performed. Its details have been explained earlier.

Form 3–Statement and Undertaking Under Section 8

From 3 is used to furnish information or actions relating to patent applications filed in other countries for the current invention. Additionally, any information relating to the rights corresponding to the present patent application has to be furnished. Further, it is used to underwrite that the applicant will keep the patent office informed in writing if he files the patent outside India.

Form 5–Declaration as to Inventorship

This application is used to declare that the inventor (s) of the subject matter ought to be protected using the current patent application.

Form 9–Request for Publication

If this form is not filed, then the patent specification will be published by the patent office after 18 months from the priority date (the filing of the first patent application for the current subject matter). On the other hand, by filling out this form, the applicant generally gets his specification published within 1 week of filling it out. It should also be noted that the patent rights start from the date of publication of the patent application (enforceable after the grant of a patent).

Form 18–Request for Examination of Application for Patent

This form can be filed within 48 months of the priority date. The patent application is not considered for examination unless this form is filed. Therefore, to expedite the patenting process, forms 9 and 18 should be filed as early as possible.

The Scope of Patenting

The scope of the patent is too great. One would think that the patent for a completely new type of "engine" would have a very broad scope. A patent for an improvement on one item of that engine may be waived to a limited extent. Patent policies encourage innovation, investment in research and development, disclosure of information, and help to problem-solve with inventions that benefit a large number of people.

Patent laws are protective, and therefore, each country has different patent laws. The Indian patent office protects only inventions filed in India. The scope of patent protection should be determined by the claims, and the claims can be determined based on specifications such as descriptions and drawings. Patents can be filed for technical work in all branches of science and technology (from basic to applied). However, it is still an emerging field that needs to be explored.

The Purpose and Advantages of Patent Laws

The main objectives of the patent law are to protect the patent (meaning the invention) and encourage the development of new technology and industry for the economic growth and development of a nation. In the absence of such protecting laws, the inventors would conceal their research work. As a result, society would remain deprived of the benefits of these inventions. As a result, these laws are extremely important and necessary in order to encourage inventors and the growth of the nation as a whole. Moreover, the inventor will be interested in disclosing the details of his invention only when he is rewarded, not otherwise.

Therefore, the major purposes of the patent laws are the following:

- To encourage researchers to make inventions or subsequent innovative work that can be practically useful to society.
- To enhance the economic growth of the community and of the nation as a whole. This can be done by granting patent protection.

The World Trade Organization (WTO) and the TRIPs agreement have increased the importance of patents around the world. Consequently, it encouraged the establishment of industries, which in turn improved the existing industries and increased employment opportunities. Motivation is very much required in the field of research, as most inventions are the result of extensive and expensive research and development. Therefore, returns on investments need to be accelerated and financial risk reduced. The patent grants the inventor certain specific advantages, which are as follows:

- It encourages the inventor to attempt innovation in research.
- It induces commercialization of the initial inventions that would otherwise have limited commercial value.
- It imparts incentives, rewards, or royalty in monetary terms for their technical innovations.
- It does not compel the inventor to file a patent for his invention under the law. Although the government encourages the disclosure of the invention for the benefit of the country, he can still keep it secret.

- It provides that the inventor can transfer, assign, or license the patent for his invention to others on payment of fees or royalties.
- It protects the invention from imitation and infringement. If the inventor fails to file a patent for his invention and suffers from pilferage or infringement, he will not be entitled to any legal remedy available to a patentee. There will always be a risk that any competitor might get the same invention patented and sue the original inventor on the grounds of infringement.
- It encourages wider use and licensing of innovative work rather than relying on secrecy.
- It forces the exposure of alternative designs and methods to make patentable products that will help in the technological growth of the industry (for example, the pharmaceutical industry).

Design

According to the Design Act 2000, "design" means the shape, texture, pattern, ornament, or composition of lines or colours applicable to any object, be it two-dimensional, three-dimensional, or two forms of an industrial process or medium, manual, mechanical, or chemical, separate or combined. It is what attracts the eye in the finished article and is completely determined by the eye. It does not contain the manufacturing process, any mechanical device, or any trademarks as defined in clause (v) of subsection (1) of Section 2 of the Trade and Commodities Act. The Design Act 2000 has undergone two amendments till now: one in 2001 and the other in 2008.

To obtain legal protection, a design must have a three-dimensional or one-dimensional shape that is two-dimensional, and the shape or pattern must apply to an article. Design is a basic requirement for safety. The basic requirement for the protection of design is that it should be novel and have some individual character and originality. Design is an additional category of patent. A design patent protects only the appearance of an article, not its structural or functional features. An example of design is the shape and design of sunglasses, pencils (cross-section: triangular, circular, hexagonal, *etc.*), water bottles, soft drink bottles (Coke, Pepsi, Kingfisher, *etc.*), vacuum flasks, the shapes of mugs and cups, pin holders, *etc.*

Designs are considered to be independent. The designs of two apparently unrelated articles, like a pair of shoes and a door handle, are claimed in separate applications. For related articles, the design can be considered distinct if it possesses different shapes and appearances. To qualify for the design, the subject must conform to the following features:

- It must have ornamental or aesthetic aspects that are useful.
- It must have a definite shape, pattern, or colour combination and
- It should be industrially reproducible (otherwise, a work of art, such as a painting, is copyrightable).

Trademarks ™

A trademark is a visual symbol in the form of a word, name, symbol, or device that is used to denote the product or goods in trade. In other words, a trademark enables a customer to differentiate the goods of one company from those of others. When properly advertised, the mark becomes an effective instrument and attracts customers by its brand name. The proper use of a trademark acquires goodwill from the clientele also. In this competitive market, the goodwill and reputation of the company are essential. It helps in developing brand strategies and builds consumer loyalty. It can also be a valuable asset to the business, which can be either sold or licensed.

The Indian Trademark Act came into force on September 15, 1999. India has taken steps to fulfil its international obligations. As a result, Indian trademark law has now become fully compliant with the international standards outlined in the TRIPS Agreement. The new law will primarily consolidate and amend the old Commercial and Commercial Marks Act 1958 and provide for better protection of goods and services.

The essential criteria for securing a trademark registration are its distinctiveness and non-deceptiveness. The Trademark Act 1999 defines it as a mark capable of being represented graphically, capable of distinguishing goods, and may include the shape of goods, their packing, and combination of colours. The other features of the trademark are as follows:

- It should be, preferably, an invented word.
- It should be easy to pronounce and remember.
- It should be short.

The trademark may include a device, brand heading, label, ticket, name, signature, word, letter, numeral, or any combination thereof. But the registration fails in the event that the following happens:

- It is likely to deceive or cause confusion among the people.
- It is merely the combination of two words or similar words.
- It is likely to hurt religious sentiments.
- It is an official seal or emblem of a country.

- It is the name of any UN organization
- It is commonly used to replace the actual name of a product with adjectives as a prefix or suffix (*e.g.*, best paint).

There are two terms, "Trademark ™" and "Registered Trademark ®", which differ from one another minutely in a way that a trademark is an unofficial registration, while a registered trademark is an official permission to make use of the trademark with its legal protection. Any new product with a distinctive and unique name can be considered to be trademarked. For example, a company can put out a new line of cricket bats called "champion", for example, complete with a graphic of a player. The graphic and the name "champion" would be considered a trademark, and the company could put the ™ designation on it immediately. A registered trademark is an official registration with the Trademark Office. A trademark (™) may be in the process of becoming registered or it may never be officially registered at all.

The term of trademark registration is 10 years. The renewal of the trademark is also possible for a further period of 10 years. Unlike patents, copyrights, or industrial designs, trademark rights can last indefinitely if the owner continues to use the mark. However, if a registered trademark is not renewed, it is liable to be removed from the register.

Trade Secret (ts)

A trade secret is a method of protecting intellectual property as a secret. It is used to describe confidential information relating to trade and commerce. It is the legal term used for confidential business information. It can be a closely guarded secret related to a process. There are certain conditions that must be fulfilled for any information to be considered a trade secret, which are as follows:

- The information must not be generally known or readily ascertainable through proper means, *i.e.*, it should not be available by obvious means.
- Due to its secrecy, the information must have independent economic value. For example, KFC is famous for its chicken recipes, which are secret.
- To protect the secrecy of the information, a trade secret holder must use reasonable measures.

A trade secret may consist of any formula, a pattern, a physical device, an idea, a process of manufacturing an article or food, *etc.* For example, the formula for preparing a soft drink, recipes, marketing strategies, manufacturing techniques, computer algorithms, and an invention for which no patent application has been filed yet. Unlike patents, trade secrets are protected without registration, *i.e.*, trade secrets are protected without any procedural formalities. Consequently, a trade secret can be protected for an unlimited time. If a trade secret is well protected, there is no definite term of protection. It can be protected for any length of time.

For example, Coke's formula is considered to be one of the best protected trade secrets. The protection lasts as long as the information is kept confidential.

It has several advantages and disadvantages. The advantage is that there is no defined term of protection and it is recommended if anyone can manage to keep the process of formulation secret, while the disadvantage is that any trade that could be discovered by "reverse engineering" cannot be protected and care must be constantly exercised to ensure confidentiality.

Domain Name

A domain name is used for an internet protocol (IP) address, which can be viewed by typing in the domain name allocated for it, *e.g.*, www.greenasia.in. The name chosen should be unique and not similar to an existing name. The main domain name is called the second level name, while the last part is called the top-level domain. For instance, "Greenasia" is a second-level name and "in" is the top-level domain.

With the advent of the internet, companies with similar names can be searched with ease, which was difficult earlier. So two companies with similar names can now come into contact, which they could not do before. Because the owner of a domain, such as Expedia.com®, provides services through the website, it is registered and protected as a trademark. Google, eBay, and Amazon are all domains turned into famous trademarks because their owners not only registered the term as a domain, but chose a term that could also be distinctive for trademark purposes.

Companies invest huge amounts of money in developing and promoting a website and not getting the website registered can prove to be fatal, as in the case of amazonnetworks.com, which dealt in computer services and had not registered its domain name as a trademark, and was sued later by amazon.com, a registered trademark domain name, which sold only books initially. Thus, it is now expedient for a domain name applicant to not only get a domain registration but also protect it as a trademark for the goods and services provided by the website.

The domain name must be unique. Under the trademark registration system, there can be a multitude of identical trademarks coexisting in the register. For example, they can be used and registered for different goods or services or used as and registered in different territories. Only one of all the proprietors that own the identical trademark can register and use the corresponding domain name. There are several proprietors who use the same trademark, Sterling, for different goods and services. Only one of them would be allowed to register the domain name sterling.co.za in South Africa. It requires renewal every year.

Geographical Indications

Geographical indications (GIs) are a class of intellectual property used to identify goods as originating in a particular territory of a country, region, or locality in that territory. The product/goods are identified by their quality, reputation, or other characteristics that are attributed to their geographical origin. The geographical indications of goods (Registration and Protection Act 1999) can be registered and protected by geographical indications in India. The TRIP's agreement requires the member countries to enact legislation for the protection of geographical indications. An office for the registration of geographical indications in India has been opened in Chennai, which is under the Office of the Controller General of Patents, Designs, and Trade Marks.

Every region has its own claim to fame. By its geographical indications, India is also best known for mangoes, tea, oranges, *etc.* Some of the examples of GI are Darjeeling tea, Kanchipuram silk sarees, Alphonso mangoes, Nagpur oranges, Kolhapuri chappals, Bikaneri bhujia, Agra ka petha, Basmati rice, *etc.*

The GIs are an indication of:

- It originates from a definite geographical territory,
- It is used to identify agricultural, natural, or manufactured goods,
- Such goods are produced, processed, or prepared within that territory and
- It has a special quality, reputation, or other characteristics.

It has several advantages. It motivates the producers to export their goods outside the territory and promotes the economic prosperity of producers in that geographical territory, and it prevents the unauthorized use of a registered geographical indication by others. It generates revenue for the growth of the territory, the state, and the country. The growth of geographical indications is a gradual process that results due to the perfect combination of nature and the skills of man and is transferred from generation to generation. Geographical indications are different from trademarks. The former denotes the identity of goods having special characteristics originating from a definite territory, whereas the latter is a sign or mark of goods or services that differs from one entrepreneur to another. So, trademarks identify a product related to a company or brand, while GI identifies a product with a particular territory.

The wines from the "Champagne" region of France are the best example; its copyright protects and prevents others from using the word "Champagne" for English wine. Even the name Champagne is not allowed for shampoo or perfume, as customers may get confused and believe that the product originated from France.

Copyright

According to the Oxford Dictionary, the word "copyright" means "copy of words. According to the Copyright Act 1957, copyright means the exclusive right to do or authorize others to do certain acts concerning literary, dramatic, or musical works or artistic works, cinematographic films, or sound recordings, *etc.*

It refers to laws that regulate the utility of the work done by the creator, such as an artist or an author. The regulations of the work include copying, distributing, altering, and displaying creative, literary, and other types of work. Unless otherwise stated in a contract, the author or creator of a work retains the copyright. It is indicated by the symbol "©". However, for copyright to apply to a work, it must be an original idea that is put to use. The idea alone cannot be protected by copyright, but it should be physically presented in the form of a written novel or poem so that it can be covered under copyright law. Copyright is a proprietary right that comes into existence as soon as the work is created. In the early days, the concept of copyright had its origins under the common law. Subsequently, it came to be governed by the statutory laws of each country. The main criterion for copyright registration is its originality.

India has a very strong and comprehensive copyright law based on the Indian Copyright Act 1957, which was amended several times, *i.e.*, in 1981, 1984, 1992, 1994, and 1999 (with effect from 15 January 2000). The 1994 Amendment was made in response to technological changes in the means of communication like broadcasting and telecasting and the emergence of new technology like computer software, while the 1999 Amendments have made the Copyright Act fully compatible with TRIP's Agreement and fully reflects the Berne Convention. The amended law provisions are for the first time to protect performers' rights as envisaged in the Rome convention. With these amendments, the Indian copyright law has become one of the most modern copyright laws in the world.

The Indian Copyright Act is valid only within the borders of the country. To secure protection for Indian works in foreign countries, India has become a member of the following international conventions on copyright and other related rights, and the copyright works of the countries mentioned in the International Copyright Order are also protected in India just like Indian works:

- The Berne Convention for the Protection of Literary and Artistic Works.
- The Universal Copyright Convention.
- The Convention for the Protection of Producers of Phonograms Against Unauthorized Duplications of Their Phonograms.
- Multilateral Conventions for the Avoidance of Double Taxation of Copyright Royalties.
- Trade Related Aspects of Intellectual Property Rights (TRIPs) Agreement.

The Indian Copyright Act protects the following works of human intellect:

1. Literary works, which include
   - Stories, plays, poems, novels, *etc.*
   - Maintenance and instruction manuals
   - Published edition of works
2. Artistic works, which include
   - Works of fine art: paintings, sculptures, drawings, diagrams, cartoons, maps, charts, *etc.*
   - Photographs, engravings
   - Cinematographic films
3. Dramatic works, which include
   - Recitation
   - Choreographic work
   - Dumb show
   - Acting
4. Computer programs
5. Electronic databases
6. Compilation work, which includes
   - Directories, who is who, Tambola ticket booklet
7. Letters: both formal and informal, *e.g.*, a letter to newspaper editors
8. Exam question paper (if assigned through contract)
9. Questionnaire for data collection
10. Research thesis and dissertation

OTHER RELATED OR NEIGHBOURING RIGHTS:

- Performers' rights
- Sound recordings
- Broadcasting rights

The Author's Right

"Author" here means

- In relation to a literary or dramatic work, the author of the work.
- In relation to a musical work, the composer.
- In relation to an artistic work other than a photograph, the artist.
- In relation to a photograph, the person taking the photograph.
- In relation to a cinematograph, sound recording, or the producer.
- In relation to any literary, dramatic, musical, or artistic work that is computer-generated, the person who causes the work to be created.

Performer's Right

As per the Indian Copyright Act, a performer includes an actor, singer, musician, dancer, acrobat, juggler, conjurer, snake charmer, person delivering a lecture, or any other person who makes a performance. 'Performance' concerning a performer's rights means any visual presentation made live by one or more performers. The performer's rights exist for 25 years.

- Right to make a sound recording or visual recording of the performance.
- Right to reproduce the sound recording or visual recording of the performance.
- Right to broadcast the performance.
- Right to communicate the performance to the public.

Broadcasting Rights

Broadcasting means communication with the public. The term of protection for a broadcaster's rights is 25 years. Certain rights are enjoyed by the broadcasting organization related to broadcasts, which are as follows:

- The right to make any sound or visual recordings of the broadcast.
- The right to make any reproduction of such a sound recording or visual recording.

WHY ARE IPR'S PROTECTED?

1. **Encourage and reward creative work:** The main social purpose of the protection of copyright and related rights is to encourage and reward creative work. This is also relevant to protection in other areas (*e.g.* industrial designs and patents).
2. **Technological innovation:** Intellectual property rights are designed to protect the results of investment in the development of new technology, thus giving the incentive and means to finance research and development activities.
3. **Fair competition:** The protection of distinctive signs and other IPRs aims to stimulate and ensure fair competition among producers.

4. **Consumer protection:** The protection of distinctive signs should also protect consumers by enabling them to make informed choices between various goods and services.
5. **Transfer of technology:** A functioning intellectual property regime should also facilitate the transfer of technology in the form of foreign direct investment, joint ventures, and licensing.
6. **Balance of rights and obligations:** It should also be noted that the exclusive rights given to the owners of intellectual property are generally subject to a number of limitations and exceptions aimed at balancing the legitimate interests of both the right holders and users.

TRIPS, TRADE AND BUSINESS: A PROVISIONAL ASSESSMENT

After providing some details on the organizational developments in the international IPR sectors, including TRIPS, and a summary review of the relevant financial analysis discussed above, we will now clarify the provisional assessment of TRIPS from a financial perspective. In short, the following discussion will focus on some selected and important questions that come up in this context.

ARE IPRS TRADE RELATED?

This question is usually recommended by the carefully chosen prefix "trade related" that first rationalized introducing IPRs into the WTO. Of course, virtually any economic regulation and/or institution will have (perhaps indirect) effects on trade, and, the other way around, trade does affect the workings of specific regulations and/or institutions. As the foregoing discussion has illustrated, it's quite apparent that weak or non-existent IPRs can affect trade in an immediate and nontrivial manner. This is often most visible for goods that are easily copied ('pirated,' in the jargon of the industries concerned), such as computer software and show business optical media products (*e.g.*, music and movies). Goods that believe in trademark protection are also quite susceptible to weak IPR protection. Indeed, firms in virtually every industry (*e.g.*, apparel, computer software and hardware, equipment, prepared food, beverages, and pharmaceuticals) depend upon trademarks for their marketing activities. Establishing a firm's reputation for quality, which may be efficiently conveyed to the buyer by known trademarks, requires significant investments in design, production, and marketing. Such costs are not borne by producers of knockoff copies or counterfeit merchandise, who, in the absence of IPRs, could easily free ride on the efforts of others. This will cause losses to the legitimate mark owners and weaken their incentive to take a position on quality production, which may be harmful to consumers (who find themselves facing a generalized lemon problem). In fact, it seems that counterfeiting and pirating have increased significantly in recent years, especially in emerging markets like China and countries of the former Soviet Union, and threaten to become a very global business [1].

It is unclear how powerful IPRs are and how they affect the quality and direction of trade. Additional trade may have an effect, as official products from the new country export local illegal copies and/or copying. However, small trade can also lead, at least for two reasons: because of incentives for IPR owners to limit production (the result of individual governance) and because strong IPRs can make legitimate domestic production [perhaps through direct foreign investment (FDI)] that is the perfect product base. Although strong evidence of a different market expansion and market power effects is unstoppable. Maskus and Penubarti [2] concluded that increased copyright protection has a positive impact on exports to OECD countries from large and small developing countries (with an impact on imports from major developing countries). Fink and Primo Braga [3] presented results that are inconsistent with these findings. Using a dynamic model for the dual-world trade of 89 countries and relying on the Ginarte and Park [4] index to measure foreign diversity in IPRs, they found a good link between IPR protection and the flow of non-fuel trading aggregates. Smith [5] analyzed exports to other American countries. It was found that copyright infringement could have a negative impact on foreign exports to the US, especially in countries that are more aggressive in imitation. Recently, Smith [6] analyzed the powerful impact of IPR exports to the US on three separate drug industries. Strong foreign IPRs tend to increase exports to the US in high-imitation countries (like other developed countries), but in countries with weak simulation capabilities, the effect of strong IPRs may be to reduce trade due to improved market forces. In any case, whether trade is positively or negatively affected by strong IPRs is not a valid question from a global economic perspective.

The issue of agriculture is of interest here because TRIPS requires little similarity of IPRs in this sector. For example, new transgenic species may be patented (and exist) in the United States, but elsewhere they may enjoy fragile protections under plant breeders' rights (PBRs). An example of Roundup Ready (RR) soybeans is instructive. In the United States, RR technology is patented, and RR soybean seeds are sold at a high premium (about 40% of the market price of common varieties) and with contractual obligations that prevent farmers from saving seeds. In Argentina, however, it is legal for farmers to save seed for re-planting purposes, and as a result, RR soybean technology is available at a lower cost to Argentine farmers than US farmers [7]. Such cost variations made by different IPRs have the potential to affect foreign competitive positions and certainly create tensions in the innovative world between the interests of the developing seed industry and the receptive agricultural sector [8].

IS POWERFUL UNIVERSAL IPR SECURITY ADVANTAGEOUS?

The main feature of TRIPS was that it required IPRs where they did not exist, without lowering any existing standards, thus providing a higher level of international protection. There is no reason to assume that strong private IPRs are desirable. From a closed (integrated) economic perspective, the optimal level of IPR protection depends on the above-mentioned sales between static losses (monopoly) and strong operating gains, *i.e.*, the need for extraction compared to the spread desire. However, as discussed by Ordover [11], this trade may be a very simple one, and a well-designed IPR protection plan does not have to go hand in hand with the spread. For example, when a patent regime is weak, innovators may choose to obtain a release using methods such as trade secrets that are worse from the point of view of ensuring the spread of new information. Whether the IPR standards are too strong or not, is a matter of debate. In the United States, for example, some are concerned that copyright protection may be severely curtailed in favour of founders, which would eliminate the difference between acquisitions and inventions, make more widespread claims and reduce innovation.

A more subtle issue is whether, as market size increases, more and more countries fall below the standard or below the normal set of IPR standards, the optimal level of IPR protection implied by those standards should increase or decrease. As mentioned earlier, because the motivation for private research and development under IPRs comes from the benefits that one person can see, it is already directly affected by the growing market, and no strengthening of IPRs would be needed. Moreover, in the international context, the consideration of the strategies mentioned earlier is contagious. The temptation to move freely makes the IPRs unilaterally set lower than those of the rest of the world, and as many countries agree to set up joint IPR protection, the most common result would be to increase the level of protection. However, as a counterpoint, the Scotchmer analysis [13] has some interest. He learned about the capabilities of IPR agreements to address issues beyond the limits of the game model, where governments have access to two innovative promotional tools: IPR protection (which promotes private R & D efforts) and public spending on R & D (which reflects the public good). Both of these instruments often create free-range outsourcing, but IPR agreements deal with just one of them. Therefore, harmonizing IPR states with negotiated agreements could ultimately result in extremely high IPR standards.

ARE TRIPS VALUABLE FOR PROGRESS?

The above discussion of international technology transfer can be viewed as part of a larger question about the impact of TRIPS on development. This is perhaps the question that has drawn the attention of analysts, fueled by the apparent asymmetry in the reforms required by TRIPS and the significant differences in both developed and developing countries at the new level of IPR-protected strategies. At present, for example, patents are primarily the work of developed countries and are not essential in many developing countries. The issues involved in analyzing the impact of TRIPS on development are complex, and the literature on this subject is simply too large to be addressed without summarizing here. One of the most recent and authoritative statements on the matter is the report of the Intellectual Property Rights Commission [14], a branch appointed by the British government. Overall, it seems we are sceptical of the constructive role that strong IPRs can play in development. The report notes that TRIPS places a heavy burden on many developing countries and supports the idea that the desired level of IPR protection depends on the developmental stage; at earlier levels, weaker (not stronger) IPRs are more likely to promote economic development. The commission also stressed that the developing world is not a single party, that a good IPR (development perspective) will inevitably differ from country to country, and that persistent IPR alignment could be detrimental to development. The report also notes that TRIPS gives developing countries greater freedom to implement the standards of IPR mandated by the treaty. However, they do not pay much attention to the main economic problems that are being emphasized by the theoretical work in this area: the unrestricted things associated with the production of new products and the problem of free commuter rides. Making claims about what the appropriate IPR policy for a developing country is exempting from this strategic problem is not enough, to say the least. After all, any small country can benefit from free rides in the new ventures of others. In fact, the problem of free commuters in this context is very broad, as reviewed by Yang [15]. Developing countries also have the motivation to travel freely with each other (in addition to free travel in developed countries), which is particularly detrimental to building technology that best suits their needs. Therefore, the choice may not be to get a new product for free or to pay for it, as the discovery of the right new South African products, as well as the timely distribution of these and other technologies, can be taken lightly. This does not mean that any policy and/or IPR standard is good. However, the claim that some IPR policies are suitable for the North and a very different approach is appropriate for the South is only one (not allowed) thesis. Another possibility is that those bad IPR policies are likely to be bad in the North and South, and what is good in one region (in terms of solving real market failures and encouraging innovation) would be good for all regions in a partnership. The (often held) view that solid IPRs are not ready for development can be very convincing if we know what is good for development.

Easterly [16] provided a sobering reminder of the failure of many past and present policies aimed at fostering growth in developing countries and ending what they call "fancy development aid myths". Learned mainly about inefficiencies: state planning, defense policies aimed at importation, price control, debt forgiveness, and allowing current use in addition to investment. As Wacziarg [17] noted, "Domestic and political policies, not international ones or capitalists, are the ones accused of tax evasion and non-productivity." While it is absurd to assume that strong private IPRs will generate greater profits for developing countries, they are likely to be an important part of the package that protects lost policies that failed in the past. Maskus [18] suggested that IPR protection in developing countries should be aligned with policies that promote strong competition and technological change, such as those aimed at liberating trade and investment, preventing corruption, promoting human capital and technical skills, and promoting social and economic freedom. Perhaps the best impact TRIPS can have on development over time, is to contribute to the establishment of political and social institutions that allow markets to operate. Focusing on agriculture, Perrin [19] noted that, without strong IPRs, it is unlikely that productivity levels in developing countries could begin to meet those of developed countries.

PRESENT ISSUES AND SCENARIO

Public Health

TRIPS came into prominence during the Doha WTO ministerial meeting in November 2001, which led to the Doha Declaration on the TRIPS and Public Health Agreement. The concerted efforts of developing countries, aided by the emerging influence of non-governmental organizations (NGOs), brought to light public health problems that plagued many of them, especially those associated with HIV/AIDS, tuberculosis, malaria, and other epidemics. The proclamation emphasized that TRIPS "does not" and should not prevent members from taking steps to protect public health. Specifically, the proclamation (a) recognized that compulsory licensing may be used to obtain sensitive drugs at the discretion of each member, (b) agreed that each member is free to use the expected form of expulsion of IPRs (subject to the corresponding import permit or not), (c) recognized that developing countries with sufficient drug production capacity would face difficulties in enforcing compulsory licenses, and thereby instructed the TRIPS Council to find a solution to this problem, and (d) extended until January 2016 the deadline for least developed countries (LDCs) to use IPR drug protection. treatment and diagnostic data. The problem referred to in point (c) above arises because TRIPS stipulates that compulsory licenses can be used primarily to provide a domestic market. The December 2002 compromise solution, which allows LDCs to cancel compulsory licenses and imports, was held by the United States in response to drug companies fears that the program could be abused, but the agreement was reached in August 2003, the day before the

Cancun meeting. The scale of the public health crisis facing many LDCs, especially in Africa, is enormous. In some countries in sub-Saharan Africa, for example, a third of adults are infected with the HIV/AIDS virus. The availability of drugs and healthcare services is a real problem. Patents on drugs, with their impact on high drug prices, can contribute to the problem, although they may not be a major stumbling block. Many important medicines, in developed and developing countries alike, do not have a patent. Poverty, lack of health insurance, and a lack of a reliable public health care system can be a real root cause of health care disasters in the third world [20]. While availability can be a major drug problem already in the community, many analysts think Doha's TRIPS efforts in the area could make a positive contribution.

STRATEGIES FOR SUCCESSFUL IPR MANAGEMENT

Effective management of intellectual property requires the implementation of a comprehensive asset management system. In this process, one of the most important steps is to review existing intellectual property to identify and obtain key company assets such as patents, trademarks, designs, trade secrets, domain names, active masks, inventions, writing tasks, hardware, and devices, depending on the type of business. Where intellectual property is not identified, it is important to determine the nature and scope of the company's intellectual property rights, which can range from direct ownership to a license, including intellectual property rights that should be developed in the future. Investing in an identified asset identified in this way requires a more constructive approach to looking at, among other things, the type of intellectual property, and the type of business that seeks intellectual property, the long-term and short-term goals of a business entity, including the intended/potential use of intellectual property.

CONCLUDING REMARKS

Growing global economic activity targets the production of goods and services that require significant R&D investment, and the exchange of these goods and services depends on the possibility of protecting the basic R&D investment in adoption by copying and imitation. Such concerns have long been voiced by the legal institutions of many developed countries, but when goods and services are sold across national borders, national IPRs are clearly inadequate. The TRIPS agreement represents the tremendous effort to find a date for these realities. An important feature of TRIPS, compared to previous IPR cooperation agreements such as those held by WIPO, is that it sets a comprehensive set of mandatory IPR protection standards for WTO membership *i.e.*, the need to continue to benefit from free trade made possible through the efforts of the 20th century GATT. In doing so, it has put the WTO into a new context, possibly suggesting a greater potential for the future of the WTO in unfamiliar areas (such as environmental standards, labour standards, competition policy, investment, and government procurement). As with any performance of this scale and scope, TRIPS is an incomplete compromise. Most importantly, TRIPS is a work in progress.

A sensitive issue in the absence of IPRs is coercion. The WTO dispute resolution process will have to deal with the parties' compliance with the treaty. The record so far provides some reason for optimism. Interestingly, most of the TRIPS disputes have so far involved developed countries, not developed countries fighting developing countries. This fact contradicts the ambiguous definition that sees TRIPS, especially as placing the interests of developed countries compared to those of developing countries. It is also helpful to note that the WTO dispute resolution mechanism provides more guarantees of provocation than the possible cases of both trade penalties under the US Special 301 process. Therefore, the WTO forum may be more attractive to developing countries facing an aggressive pro-agenda agenda. However, the problems with TRIPS enforcement are likely to be at the national level, following the acquisition of TRIPS compliance. It remains to be seen how the effective national IPR protection legislation will work, especially in developing countries, and how TRIPS' illegal economic behaviour will be enforced (definitely incomplete). In addition to the remaining challenges of implementation, there are opportunities to expand and improve the TRIPS agreement. What is doubtful is that any agreement may simply be given to the evolving goals of the developing countries. Although developed countries can not be considered to have a united agenda, the most important thing they would like to strengthen is the existing TRIPS, the closing of the gaps, and the expansion of the protection zone under the treaty.

TRIPS offers great flexibility in many areas, especially in IPRs related to new technologies. No provision has been made for TRIPS, for example, regarding IPRs related to Internet data transfer and e-commerce. Indeed, two agreements in this area have been finalized under WIPO following the termination of the Uruguay Round (Copyright Treaty and Performances and Phonogram Treaties), and developed countries would like to see greater provision for these agreements brought into TRIPS. There is also an interest in specifying patent protection for new biotechnology products, and perhaps even in a review of the provision that allows members to exclude plants and animals from the patent (in fact, TRIPS is considering mandatory reviews built within this section of the law). At the very least, the United States would like to see the UPOV 1991 conference clearly identified as the level of protection of *sui generis* for plant species. Developing countries, on the other hand, have an agenda that is almost one of many problems and aims to protect the flexibility provided by TRIPS (including the freedom of choice of sui generis system for more fragile plant species than UPOV 1991, especially concerning farmers' right to save and exchange seeds). Developing countries would also like to reconcile the provisions of TRIPS and the CBD, including, as discussed earlier, the issue of whether disclosure requirements for the use of biodiversity sources may be required. The integration of the agendas of developed countries with TRIPS should not be overlooked, as differences exist in many areas. One ongoing bone of contention is the reliance on the US in the first patent law, as opposed to the first file system used by almost all other WTO members. The European Union and the United States are still embroiled in controversy over plant and animal ownership. As mentioned earlier, the issue of GIs is causing European countries and many developing countries to run against the United States and other developed countries of the new world. As evidenced by their presence in debates on the impact of TRIPS on access to essential medicines in poor countries, non-governmental organizations will also be instrumental in making amendments to TRIPS. The growing frustration of developing countries over allegations that the Uruguay Round quid pro quo (TRIPS in exchange for textile and agricultural permits) also suggests that the benefits and costs of the TRIPS transformation may need to be traded for its benefits rather than relying on compensation for the opposite agreement.

CONSENT FOR PUBLICATION

Not applicable.

CONFLICT OF INTEREST

The author declares no conflict of interest, financial or otherwise.

ACKNOWLEDGEMENTS

Declared none.

REFERENCES

[1]   Imitating Property is Theft. 2003. https://www.economist.com/special-report/2003/05/15/imitatin-
       -property-is-theft

[2]   Maskus KE, Penubarti M. How trade-related are intellectual property rights? J Int Econ 1995; 39: 227-
       48.

[3]   Fink C, Braga CAP, Eds. How Stronger Protection of Intellectual Property Rights Affects International
       Trade Flows. Washington, DC: The World Bank 1999; p. 23.

[4]   Ginarte JC, Park W. Determinants of intellectual property rights: a cross national study. Res Policy
       1997; 26: 283-301.

[5]   Smith PJ. Are patent rights a barrier to US exports? J Int Econ 1999; 48: 151-77.

[6]   Smith PJ. Patent rights and trade: analysis of biological products, medicinals and botanicals, and
       pharmaceuticals. Am J Agric Econ 2002; 84: 495-512.

[7]   Biotechnology: Information on Prices of Genetically Modified Seeds in the United States and Argentina. Washington, DC: USA 2000; p. 25. https://www.gao.gov/assets/230/228726.pdf

[8]   Moschini G, Lapan H, Sobolevsky A. Roundup ready soybeans and welfare effects in the soybean complex. Agribusiness 2000; 16: 33-55.

[9]   Arup C, Ed. The New World Trade Organization Agreements: Globalizing Law Through Services and Intellectual Property. Cambridge University Press 2000; p. 336.

[10]   Maskus KE, Chen Y. Parallel imports in a model of vertical distribution: theory, evidence, and policy. Pac Econ Rev 2002; 7: 319-34.

[11]   Ordover JA. A patent system for both diffusion and exclusion'. J Econ Perspect 1991; 5: 43-60.

[12]   Merges RP. As many as six impossible patents before breakfast: property rights for business concepts and patent system reform. Berkeley Technol Law J 1999; 14: 577-615.

[13]   Scotchmer S, Ed. The political economy of intellectual property treaties. Compt Policy Cent. USA: Univ Berkeley 2002; p. 30.

[14]   Integrating Intellectual Property Rights and Development Policy. London, UK: Commission on Intellectual Property Rights 2003; p. 190.

[15]   Yang Y. Why do southern countries have little incentive to protect northern IPRs? Can J Econ 1998; 31: 800-16.

[16]   Easterly WR, Ed. The Elusive Quest for Growth: Economists' Adventures and Misadventures in the Tropics. Cambridge, MA: MIT Press 2001; p. 356.

[17]   Wacziarg R. Review of Easterley's the elusive quest for growth. J Econ Lit 2002; 40: 907-18.

[18]   Maskus KE, Ed. Intellectual Property Rights in the Global Economy. Washington, DC: Peterson Institute for International Economics 2000; p. 296.

[19]   Perrin RK. Intellectual property rights and developing country agriculture. Agric Econ 1999; 21: 221-9.

[20]   Anania G, Bohman ME, Carter CA, *et al.* Eds Agricultural Policy Reform and WTO: Where Are We Heading?. Cheltenham, UK: Edward Elgar Publishing 2004; p. 672.

SUBJECT INDEX

A

Abiotic stress tolerance 11
ACC deaminase activity 11
Acetobacter aceti 72
Acetyl-CoA 50
Acholeplasma 167
Achromobacteriaceae 14
Acidification 93, 215
Acidithiobacillus ferrooxidans 73, 143
Acidophiles 3, 4
Acid(s) 15, 17, 24, 40, 41, 42, 43, 46, 48, 49,
    50, 51, 52, 59, 65, 67, 71, 72, 107, 109,
    116, 129, 130, 142, 143, 153, 154, 155,
    156, 157, 160, 168, 172, 173, 182, 221
  abscisic 109
  acetic 50, 51, 71, 72
  amino (AAs) 15, 40, 42, 71, 107, 116, 154,
    155, 160, 172, 182
  ascorbic 72
  butyric 49, 50, 51, 71
  carbonic 156
  citric 41, 43, 67, 72, 156
  ethylenediaminetetraacetic 143
  gibberellic 65, 67
  gluconic 156
  glycolic 17
  inorganic 156, 168, 172
  lactic 48, 49, 50, 52, 71, 72
  naphthenic 142
  nicotinic 65
  nitric 156
  nucleic 41, 46, 154, 155, 221
  organic 41, 43, 59, 65, 67, 72, 129, 130,
    156, 157, 160, 168, 173
  oxalic 156
  pantothenic 65
  propionic 71, 72
  pseudomonic 24
  shikimic chorismic 52
  succinic 49, 50, 156
  tartaric 156

Actinobacteria 1, 8, 9, 109, 158, 226
Activated carbon adsorption processes 125
Activation 5, 25, 140, 228
  degradative enzyme 160
  energies 223
Activities 1, 9, 21, 24, 27, 28, 52, 61, 65, 73,
    92, 107, 134, 138, 139, 140, 144, 162,
    169, 171, 211, 216, 217, 241
  agricultural 92
  antagonistic 9, 107
  antibacterial 52
  antifungal 73
  anti-fungal 27
  antimicrobial 24
  antinematode 24
  broad-spectrum 169, 171
  enzymatic 28, 140, 216, 217
  metabolic 21, 61, 65, 134, 144, 162, 211
  soil biota 1
Acts 2, 5, 90, 94, 99, 107, 134, 138, 139, 160,
    144, 241
  ligninolytic enzymes 160
Adhesion 48, 70, 224
  interfacial 224
Adsorption 17, 125, 141, 158, 221, 222
  efficient 222
Aerobacter aerogenes 126, 127
Aerobic catabolism 22, 23
Agents 24, 48, 53, 60, 65, 66, 70, 73, 115,
    121, 145, 163, 170, 172, 226, 242
  antifungal 53, 65
  antimicrobial 73, 226
  anti-parasitic 60
  bio-pulping 172
  bulking 145
  cytotoxic 66
  fungicidal 170
  immobilizing 70
  infectious 115
  lytic 24
Agglomeration 221

Y

Z

www.ingramcontent.com/pod-product-compliance
Lightning Source LLC
Chambersburg PA
CBHW050814220326
41598CB00006B/205